T0205687

CHEMICAL TECHNOLOGY AND INFORMATICS IN CHEMISTRY WITH APPLICATIONS

Innovations in Physical Chemistry: Monograph Series

CHEMICAL TECHNOLOGY AND INFORMATICS IN CHEMISTRY WITH APPLICATIONS

Edited by

Alexander V. Vakhrushev, DSc
Omari V. Mukbaniani, DSc
Heru Susanto, PhD

Apple Academic Press Inc.
3333 Mistwell Crescent
Oakville, ON L6L 0A2
Canada

Apple Academic Press Inc.
9 Spinnaker Way
Waretown, NJ 08758
USA

© 2019 by Apple Academic Press, Inc.

First issued in paperback 2021

Exclusive worldwide distribution by CRC Press, a member of Taylor & Francis Group

No claim to original U.S. Government works

ISBN 13: 978-1-77463-158-4 (pbk)
ISBN 13: 978-1-77188-666-6 (hbk)

Library and Archives Canada Cataloguing in Publication

Chemical technology and informatics in chemistry with applications / edited by
Alexander V. Vakhrushev, DSc, Omari V. Mukbaniani, DSc, Heru Susanto, PhD.

(Innovations in physical chemistry. Monograph series)
Includes bibliographical references and index.
Issued in print and electronic formats.
ISBN 978-1-77188-666-6 (hardcover).--ISBN 978-1-351-24745-0 (PDF)

1. Chemistry, Technical. 2. Cheminformatics. I. Vakhrushev, Alexander V., editor II. Mukbaniani,
Omari V., editor III. Susanto, Heru, 1965-, editor IV. Series: Innovations in physical chemistry.
Monograph series

TP145.C54 2018	660	C2018-905193-0	C2018-905194-9

CIP data on file with US Library of Congress

Apple Academic Press also publishes its books in a variety of electronic formats. Some content that appears in print may not be available in electronic format. For information about Apple Academic Press products, visit our website at **www.appleacademicpress.com** and the CRC Press website at **www.crcpress.com**

ABOUT THE EDITORS

Alexander V. Vakhrushev, DSc

Alexander V. Vakhrushev, DSc, is a Professor at the M. T. Kalashnikov Izhevsk State Technical University in Izhevsk, Russia, where he teaches theory, calculating, and design of nano- and microsystems. He is also the Chief Researcher of the Department of Information-Measuring Systems of the Institute of Mechanics of the Ural Branch of the Russian Academy of Sciences and Head of the Department of Nanotechnology and Microsystems of Kalashnikov Izhevsk State Technical University. He is a corresponding member of the Russian Engineering Academy. He has over 400 publications to his name, including monographs, articles, reports, reviews, and patents. He has received several awards, including an academician A. F. Sidorov Prize from the Ural Division of the Russian Academy of Sciences for significant contribution to the creation of the theoretical fundamentals of physical processes taking place in multilevel nanosystems and honorable scientist of the Udmurt Republic. He is currently a member of editorial board of several journals, including *Computational Continuum Mechanics, Chemical Physics and Mesoscopia,* and *Nanobuild.* His research interests include multiscale mathematical modeling of physical–chemical processes into the nanohetero systems at nano-, micro-, and macro-levels; static and dynamic interaction of nanoelements; and basic laws relating the structure and macro characteristics of nanohetero structures.

Omari V. Mukbaniani, DSc

Omari Vasilii Mukbaniani, DSc, is Professor and Chair of the Macromolecular Chemistry Department of Iv. Javakhishvili Tbilisi State University, Tbilisi, Georgia. He is also the Director of the Institute of Macromolecular Chemistry at Iv. Javakhishvili Tbilisi State University. He is a member of the Academy of Natural Sciences of Georgia. For several years he was a member of advisory board and editorial board of the Journal Proceedings of Iv. Javakhishvili Tbilisi State University (Chemical Series) and contributing editor of the journals *Polymer News, Polymers Research Journal,* and *Chemistry and Chemical Technology.* His research interests include polymer chemistry, polymeric materials, and chemistry of organosilicon compounds, as well as methods of precision synthesis to build block and the development of graft and comb-type structure. He also researches the mechanisms of reactions leading to these polymers and the synthesis of various types of functionalized silicon polymers, copolymers, and block copolymers.

Heru Susanto, PhD

Heru Susanto, PhD, is currently Head of the Information Department and researcher at the Indonesian Institute of Sciences, Computational Science & IT Governance Research Group. At present he is also an Honorary Professor and Visiting Scholar at the Department of Information Management, College of Management, Tunghai University, Taichung, Taiwan. Dr. Susanto has worked as an IT professional in several roles, including Web Division Head of IT Strategic Management at Indomobil Group Corporation and Prince Muqrin Chair for Information Security Technologies at King Saud University. His research interests are in the areas of information security, IT governance, computational sciences, business process re-engineering, and e-marketing. Dr. Susanto received a BSc in Computer Science from Bogor Agricultural University, an MBA in Marketing Management from the School of Business and Management Indonesia, and an MSc in Information Systems from King Saud University, and a PhD in Information Security System from the University of Brunei and King Saud University.

INNOVATIONS IN PHYSICAL CHEMISTRY: MONOGRAPH SERIES

This book series offers a comprehensive collection of books on physical principles and mathematical techniques for majors, non-majors, and chemical engineers. Because there are many exciting new areas of research involving computational chemistry, nanomaterials, smart materials, high-performance materials, and applications of the recently discovered graphene, there can be no doubt that physical chemistry is a vitally important field. Physical chemistry is considered a daunting branch of chemistry—it is grounded in physics and mathematics and draws on quantum mechanics, thermodynamics, and statistical thermodynamics.

Editors-in-Chief

A. K. Haghi, PhD
Editor-in-Chief, *International Journal of Chemoinformatics* and *Chemical Engineering and Polymers Research Journal*; Member, Canadian Research and Development Center of Sciences and Cultures (CRDCSC), Montreal, Quebec, Canada
E-mail: AKHaghi@Yahoo.com

Lionello Pogliani, PhD
University of Valencia-Burjassot, Spain
E-mail: lionello.pogliani@uv.es

Ana Cristina Faria Ribeiro, PhD
Researcher, Department of Chemistry, University of Coimbra, Portugal
E-mail: anacfrib@ci.uc.pt

BOOKS IN THE SERIES

- **Applied Physical Chemistry with Multidisciplinary Approaches**
  Editors: A. K. Haghi, PhD, Devrim Balköse, PhD, and
  Sabu Thomas, PhD

- **Chemical Technology and Informatics in Chemistry with Applications**
  Editors: Alexander V. Vakhrushev, DSc, Omari V. Mukbaniani, DSc, and Heru Susanto, PhD

- **Engineering Technologies for Renewable and Recyclable Materials: Physical-Chemical Properties and Functional Aspects**
  Editors: Jithin Joy, Maciej Jaroszewski, PhD, Praveen K. M.,
  and Sabu Thomas, PhD, and Reza Haghi, PhD

- **Engineering Technology and Industrial Chemistry with Applications**
  Editors: Reza Haghi, PhD, and Francisco Torrens, PhD

- **High-Performance Materials and Engineered Chemistry**
  Editors: Francisco Torrens, PhD, Devrim Balköse, PhD,
  and Sabu Thomas, PhD

- **Methodologies and Applications for Analytical and Physical Chemistry**
  Editors: A. K. Haghi, PhD, Sabu Thomas, PhD, Sukanchan Palit,
  and Priyanka Main

- **Modern Physical Chemistry: Engineering Models, Materials, and Methods with Applications**
  Editors: Reza Haghi, PhD, Emili Besalú, PhD, Maciej Jaroszewski,
  PhD, Sabu Thomas, PhD, and Praveen K. M.

- **Physical Chemistry for Chemists and Chemical Engineers: Multidisciplinary Research Perspectives**
  Editors: Alexander V. Vakhrushev, DSc, Reza Haghi, PhD,
  and J. V. de Julián-Ortiz, PhD

- **Physical Chemistry for Engineering and Applied Sciences: Theoretical and Methodological Implication**
  Editors: A. K. Haghi, PhD, Cristóbal Noé Aguilar, PhD,
  Sabu Thomas, PhD, and Praveen K. M.

- **Research Methodologies and Practical Applications of Chemistry**
  Editors: Lionello Pogliani, PhD, A. K. Haghi, PhD,
  and Nazmul Islam, PhD

- **Theoretical Models and Experimental Approaches in Physical Chemistry: Research Methodology and Practical Methods**
  Editors: A. K. Haghi, PhD, Sabu Thomas, PhD, Praveen K. M.,
  and Avinash R. Pai

CONTENTS

CONTRIBUTORS

Ann Rose Abraham
School of Pure and Applied Physics, Mahatma Gandhi University, Kottayam, Kerala 686560, India

Teuku Beuna Bardant
Indonesian Institute of Science, Jakarta Timur, Indonesia

Gloria Castellano
Departamento de Ciencias Experimentales y Matemáticas, Facultad de Veterinaria y Ciencias Experimentales, Universidad Católica de Valencia San Vicente Mártir, Guillem de Castro-94, E-46001 València, Spain

Natarajan Chandrasekaran
Centre for Nanobiotechnology, VIT University, Vellore, Tamil Nadu, India

A. K. Chaudhary
Advanced Centre of Research in High Energy Materials, University of Hyderabad, Hyderabad 500046, Telangana, India

Ching Kang Chen
School of Business and Economics, Bandar Seri Begawan, Brunei

Ganesh Damarla
Advanced Centre of Research in High Energy Materials, University of Hyderabad, Hyderabad 500046, Hyderabad, India

Nandakumar Kalarikkal
School of Pure and Applied Physics, Mahatma Gandhi University, Kottayam, Kerala 686560, India

Fang-Yie Leu
Tunghai University, Taichung, Taiwan

Amitava Mukherjee
Centre for Nanobiotechnology, VIT University, Vellore, Tamil Nadu, India

Sukanchan Palit
Department of Chemical Engineering, University of Petroleum and Energy Studies, Bidholi via Premnagar, Dehradun, Uttarakhand 248007, India
43, Judges Bagan, Haridevpur, Kolkata 700082, India

Vidhya Hindu S.
Centre for Nanobiotechnology, VIT University, Vellore, Tamil Nadu, India

Arief Amier Rahman Setiawan
Indonesian Institute of Science, Jakarta Timur, Indonesia

D. S. Shuklin
Kalashnikov Izhevsk State Technical University, Izhevsk, Russia

S. G. Shuklin
Kalashnikov Izhevsk State Technical University, Izhevsk, Russia

Heru Susanto
Indonesian Institute of Science, Jakarta Timur, Indonesia
Department of Information Management, College of Management, Tunghai University,
Taichung, Taiwan

John Thomas
Centre for Nanobiotechnology, VIT University, Vellore, Tamil Nadu, India

Sabu Thomas
International and Inter University Centre for Nanoscience and Nanotechnology,
Mahatma Gandhi University, Kottayam, Kerala 686560, India
School of Chemical Sciences, Mahatma Gandhi University, Kottayam, Kerala 686560, India
International and Inter University Centre for Nanoscience and Nanotechnology,
Mahatma Gandhi University, Kottayam, Kerala 686560, India

Francisco Torrens
Institut Universitari de Ciència Molecular, Universitat de València, Edifici d'Instituts de Paterna,
P. O. Box 22085, E-46071 València, Spain

Rehana P. Ummer
School of Pure and Applied Physics, Mahatma Gandhi University, Kottayam, Kerala 686560, India

A. V. Vakhrushev
Kalashnikov Izhevsk State Technical University, Izhevsk, Russia
Institute of Mechanics, Ural Division, Russian Academy of Sciences, Izhevsk, Russia

M. Venkatesh
Advanced Centre of Research in High Energy Materials, University of Hyderabad,
Hyderabad 500046, Telangana, India

ABBREVIATIONS

ADRs	adverse drug reactions
AFM	atomic force microscopy
AIDS	acquired immune deficiency syndrome
AOP	advanced oxidation processes
AOTs	advanced oxidation technologies
BSS	bare semiconductor surfaces
CAR	chimeric antigen receptor
CFCs	chlorofluorocarbons
CNT	carbon nanotubes
CPM	composite polymer materials
DDBJ	DNA database of Japan
DWP	dual-wave plate
EA	easy axis
EC	epoxy composition
ECN	ECs with nanostructures
EDCs	endocrine-disrupting compounds
EFB	empty fruit bunch
EMR	electronic medical record
EO	electro-optic
FAFNER	factoring via network-enabled recursion
FC	field cooled
FDA	Food and Drug Administration
FE	field emission
FeRAMs	ferroelectric random-access memories
FTP	file transfer protocol
GA	genetic algorithms
GPS	global positioning system
GPU	graphics processing unit
GRAVY	grid-enabled virtual file system
HFCs	hydrofluorocarbons
HIV	human immunodeficiency virus
HPC	high-performance computing
IaaS	infrastructure-as-a-service

ICT	information and communication technology
IS	information system
IT	information technology
I-WAY	information wide area year
JRM	jump relaxation model
KEGG	Kyoto Encyclopedia of Genes and Genomes
KNIME	Konstanz Information Miner
LANL	Los Alamos National Laboratory
LCD	liquid crystal display
LGG	lactobacillus rhamnosus GG
LT-GaAs	low-temperature gallium arsenide
MCDM	multi-criteria decision-making
MC	magneto-capacitance
MD	magnetodielectric
MEA	monoethanolamide
ME	magnetoelectric
MF	multiferroics
MNTs	multilayered carbon NTs
MOEA	multi-objective evolutionary algorithms
MOO	multi-objective optimization
MRAM	magnetic memory device
NCs	nanocomposites
NF	nanofiltration
NLC	nematic liquid crystal
NP	nanoparticles
NT	nanotubes
OAI-ORE	open archives initiative object reuse and exchange
PaaS	platform-as-a-service
PANs	peroxyacyl nitrates
PCA	photoconductive antennas
PC	personal computer
PDB	Protein Data Bank
PDE	Photo-Dember effect
PMMA	polymethyl methacrylate
PPCP	pharmaceuticals and personal care products
PS3	Play Station 3
PVA	polyvinyl alcohol
QoS	quality of service

QSAR	quantitative structure–activity relationship
ReRAM	resistive random access memory
RS	resistive switching
SaaS	software-as-a-service
SAR	structure–activity relationship
SA	simulated annealing
SDF	surface depletion field
SEM	scanning electron microscope
SI	semi-insulating
SKOS	Simple Knowledge Organization System
SMB	simulated moving bed
SQL	Structured Query Language
SRF	strontium hexaferrite
STM	scanning tunneling microscopy
TCP	Transmission Control Protocol
TCR	tumor antigen receptor
TEM	transmission electron microscopy
UV	ultraviolet
VOs	virtual organizations
VSM	vibrating sample magnetometer
WWTP	wastewater treatment plant
XPS	x-ray photoelectron spectroscopy
XRD	x-ray diffraction
ZFC	zero field cooled

PREFACE

This new volume includes a variety of modern research topics in industrial and engineering chemistry research.

The book emphasizes fundamental concepts while presenting cutting-edge research developments to emphasize the vibrancy of industrial and engineering chemistry research today.

This book covers basic concepts and practical applications, focuses heavily on research, and fills a need for current information on the science, processes, and applications in the field.

It examines the industrial process for emerging materials, determines practical use under a wide range of conditions, and establishes what is needed to produce a new generation of materials.

This volume serves as a reference book for advanced/graduate students in chemical engineering and applied sciences. Introducing readers to the latest research applications, this new book:

- Studies some of the most important aspects of industrial chemistry and cheminformatics
- Focuses on varied aspects of emerging industrial chemistry and cheminformatics in a single volume
- Contains relevant examples, illustrations, and applications
- Provides qualitative and quantitative information on new techniques in chemistry and cheminformatics
- Describes the engineering aspects involved in physical chemistry
- Contains cutting-edge information on emerging technologies
- Covers new aspects of chemistry and physics, biochemistry development, and technology

PART I
New Insights in Engineering Chemistry

CHAPTER 1

USE OF PROBIOTICS AS A FEED IN AQUACULTURE: A REVIEW

VIDHYA HINDU S., NATARAJAN CHANDRASEKARAN, AMITAVA MUKHERJEE, and JOHN THOMAS*

Centre for Nanobiotechnology, VIT University, Vellore, Tamil Nadu, India, Tel.: +91 416 2202876, Fax: +91 416 2243092

*Corresponding author. E-mail: john.thomas@vit.ac.in; th_john28@yahoo.co.in

ABSTRACT

One of the rapidly developing food production sectors in India is aquaculture. For reducing the pathogen in aquaculture, antibiotics have been used as feed additives for many decades. However, the use of antibiotics in aquatic animals may cause resistance to the microorganisms. The need for efficiency of feed digestion, disease resistance, growth, and immune system enhancement of aquatic animals has increased in the aquaculture industry. The maintenance of intestinal microbes in the host gut is the important criterion for the growth of healthy aquatic animals. Probiotics are one such alternative that exerts their beneficial effect when supplemented through fish diet to the host. This chapter summarizes the beneficial application of probiotics as a feed for aquatic animals to increase the aquaculture production by enhancing growth rate, immune response, water quality, and preventing colonization of pathogen in the gastrointestinal tract.

1.1 INTRODUCTION

In many countries, aquaculture has become one of the most important economic activities. During the production on a large scale in aquaculture, aquatic animals were exposed to pathogenic diseases and so forth that resulted in a severe loss of the economy. Aquatic animals are the fastest growing sources of income due to their nutritional properties to the human beings. Diseases caused by pathogenic bacteria are responsible for causing great economic loss to aquaculture farmers. In the aquaculture industry, control of diseases has been achieved by using antibiotics. However, the antibiotics use for controlling diseases in aquaculture leads to the development of drug resistance among fish, increases the stress on aquatic animals, and also poses a potential risk to humans. The use of probiotics is an alternative method for controlling fish disease and maintaining a healthy aquatic environment. The use of probiotics in aquaculture will reduce the use of antimicrobial drugs that were prophylactic.[38]

Fish bodies contain a group of microbes; each microbe performs different functions. Microbes present in the gastrointestinal tract plays a major role in the digestion and absorption of feed ingested by the host which in turn enhances the immunity and disease resistance of the host. Due to harsh environmental conditions, an intestinal microbial community may get affected. To overcome this, use of probiotics with feed will manipulate the intestinal microbial community. Probiotics are beneficial microorganisms, which can be developed into different products with food, drugs, and dietary supplements. Ganguly[10] reported that when immunostimulants, probiotics, and prebiotics are supplemented in feed, they generally improve feed efficiency, growth performance, and the immune response of crustaceans and fishes even when administered in small quantities.

1.2 OVERVIEW OF PROBIOTICS

Gismondo et al.[15] reported that the term probiotic originated from the Greek words "pro" and "bios" that mean "for life." The probiotics are generally referred to as a supporter of life that helps to increase the health of the host organism. The effect of the probiotic use on intestinal floral balance was defined and demonstrated only in some cases.[23] The World Health Organization defines probiotic as a live microorganism which confers a health advantage to the host organism when provided in an acceptable quantity.

Fuller[9] re-improved the term "beneficial effects of live microbial probiotic feed supplement improves the intestinal balance of the host animal." Gatesoupe[12] described probiotics as "microbial cells administered in a certain way, which reaches the gastrointestinal tract with the aim of improving health." Parker[32] termed probiotic as "organisms and substances that contribute to intestinal microbial balance." He had coined this term to define a microbial feed supplement.

The US Food and Drug Administration (FDA)[45] used other terms to define live microbes which are used for regulatory function. FDA, in 1995, termed live microorganisms that are used in animal feeds as "direct fed microbial." When designed for the treatment as human drugs, they are classified as "live biotherapeutics."[19] The use of probiotics or beneficial bacteria has increased recently due to its antimicrobial effect by modifying the intestinal microbial balance. As a result, the probiotics secretes antibacterial substances such as bacteriocins and organic acids that compete with the pathogen to inhibit the attachment of pathogen to the intestine.[24]

A variety of populations of microorganisms is present in the gastrointestinal tract, and their microbial population is damaged by various agents like stress, medication, diet, age, and environment.[10] It is essential to note that the endogenous microbial population depends on many factors such as genetic, nutritional, and environmental factors. Microorganism plays a major role in the health status of aquatic species environment than in the humans and terrestrial animals. The gastrointestinal tract of aquatic animals is probably constituted by the indigenous microbiota. This microbiota jointly with artificially high levels of allochthonous microorganisms maintains the constant ingestion from the surrounding water.[18] Most commonly used bacterial genera in probiotic preparations are *Bacillus, Escherichia, Streptococcus, Lactobacillus, Bifidobacterium,* and *Enterococcus.* Some of the fungal strains belonging to *Saccharomyces* are also used in the preparation.

Preetha et al.[36] reported that probiotics have been isolated from various sources such as aquatic environment, water, and sand. Probiotics have also been isolated from skin mucus of fish.[43] They have also been isolated from the gastrointestinal tract of the aquatic animals.[29]

1.2.1 TYPES OF PROBIOTICS AND THEIR USE

There are basically three types of probiotics: (a) gut probiotics which can be administered orally by blending with feed to improve the helpful gut microorganism, (b) water probiotics which are present in water medium to eliminate the pathogenic microorganism from proliferation and also from consuming available nutrients,[28] (c) soil probiotics are used to control bottom pond pollution and also to control soil pH to prevent the prevalence of poisonous gases such as nitrogen, ammonia, methane, and so forth.

1.2.2 SELECTION OF PROBIOTICS

The use of microorganisms in aquaculture as a probiotic should exert their antimicrobial activity against pathogens and should be secure for the aquatic host and also for their enclosing environments and humans.[27] The bacterial strain should fulfill the following criteria to be a probiotic:

a) Selected bacterial strain should be safe for the host (i.e., nonpathogenic and nontoxic)
b) Should be able to produce antimicrobial substance against pathogenic bacteria
c) Should be tolerant to pH and bile acids
d) Should adhere to the intestinal epithelial cells
e) Should enhance the immune response against infectious agents
f) Should be viable and stable during storage
g) Should persist in the digestive system for a long time to exert their beneficial effect

1.2.3 MECHANISM OF ACTION OF PROBIOTICS

Several studies have been published on probiotics during the last decade. However, it is tricky to comprehend the mechanism of action of probiotics and only limited descriptions are accessible because of the methodological ethical limitation of aquatic animals.[4] The process by which the probiotic bacteria interact with gastrointestinal tract is not well understood, however, the probable mechanisms have been reported.

a) Probiotic bacteria could eliminate the pathogenic microorganism or produce antibacterial substances that reduce the development of pathogenic bacteria.
b) Probiotics provides an essential nutrient that helps to strengthen the nutrition of the cultured aquatic animals.
c) They provide digestive enzymes to improve digestion in the intestine of cultured aquatic animals.

d) Probiotics improves the quality of water by decomposing the toxic substances and organic matter in water.

e) They enhance the immune response of cultured animals against pathogenic microorganisms.

1.3 ADMINISTRATION OF PROBIOTICS THROUGH FEED FOR ANIMALS

Probiotic bacteria can directly be applied to feed. In some cases, to stabilize the probiotic bacteria in feed, binding agents such as agar, sodium alginate, cassava starch, gelatin, polyvinyl alcohol, tapioca powder, wheat gluten, and kelp meal are used.[2]

Researchers, fish farmers, and animal feed production companies have been looking forward to the alternative products that can help to reduce the entry of pathogenic microorganism into the animal gut and also for the improved growth of aquatic animals to increase their production. The utilization of probiotics in feed can be an alternative to antibiotics for growth enhancement and to improve the gut microbial balance.

Probiotics are beneficial microorganisms and supportive substances, which when ingested by humans and animals produce a constructive effect by helping in the establishment of an intestinal population which is advantageous to the host entity and antagonistic to pathogenic bacteria.[40] Prepared feed not only provides nutrients for growth and development of aquatic animals but also controls the stress and disease caused by the environment. However, probiotic-supplemented feed should possess the above criteria.

Probiotics can be administered either singly or in combination through the feed. The administration of *Lactobacillus acidophilus* or *Bifidobacterium bifidum* with feed-protected the Nile tilapia challenged with *Aeromonas hydrophila*.[3] Iwashita et al.[20] reported that a diet supplemented with a combination of probiotics of *Bacillus subtilis*, *Saccharomyces cerevisiae*, and *Aspergillus oryzae* increased respiratory burst activity, erythrocyte fragility, and levels of white blood cells of Nile tilapia.

Probiotics exist in different ways in which the bacterium will be helpful, and probiotics could act either individually or in the blend, forming a single probiotic. These include pathogen inhibition by producing antibacterial substance, competition for sites of attachment, nutrient competition, modification of enzymatic activity of pathogens, immunostimulatory functions, and nutritional benefits such as improving digestibility and utilization of feed.[21] According to Food and Agricultural Organization and World Health Organization guidelines, the use of probiotic organisms in food should have the capability of surviving in the gastrointestinal tract, that is, they should be tolerant to the acidic condition of the stomach. Furthermore, it must have the capacity to colonize and multiply in the gastrointestinal tract. They should be sufficiently protected to maintain the product shelf life effectively.[39]

1.4 ADMINISTRATION OF ENCAPSULATED PROBIOTIC THROUGH FEED

Encapsulation of probiotic bacteria is an efficient technology to enhance the probiotic viability in the gastrointestinal tract. During storage in the gastrointestinal tract, the probiotic viability is affected by adverse environmental conditions of pH, hydrogen peroxide production, temperature, and bile acid in the stomach. To protect from adverse conditions, probiotic cells covered with physical barrier is an approach that received considerable interest.

Sultana et al.[41] reported that microencapsulation can be defined as the process which reduces the cell injury by retaining the cells within an encapsulating membrane that results in the appropriate release of microorganism in the gut. Components such as alginate, starch, xanthan-gellan, carrageenan, gelatin, chitosan, miscellaneous compounds, and cellulose acetate phthalate can be used as the encapsulating membrane/polymers for the microencapsulation of probiotics.[26] Several methods have been used for the encapsulation of probiotics in the polymer. The microencapsulation process is selected based on physical and chemical properties of core and coating material.[35] Gibbs et al.[14] reported various encapsulation techniques for encapsulating probiotics such as extrusion coating, coacervation, centrifugal suspension separation, lyophilization, spray drying, spray chilling, fluidized bed coating, liposomal entrapment, cocrystallization, and inclusion complexation.

Reports suggest that microencapsulation of probiotics increases the viability when passing through the acidic–enzymatic–bile condition of the intestinal tract of aquatic animals.[26] Pirarat et al.[34] evaluated the viability of free and alginate-encapsulated *Lactobacillus rhamnosus* GG (LGG) under simulated gastrointestinal conditions of tilapia. After being exposed to simulated gastric conditions, the viability was significantly higher in encapsulated LGG than the free LGG. Pinpimai et al.[33] observed the viability and morphology of encapsulated yeast, *S. cerevisiae* during storage and passage through the stimulated gut and bile conditions. The viability was significantly higher in simulated gastric and bile conditions of encapsulated yeast compared to that of free yeast after storage at room temperature for 14 days.

In another study, probiotics such as *L. acidophilus* and *Bifidobacterium* spp. encapsulated in calcium alginate beads upgraded the probiotic survival in yogurt at 5°C during 8 weeks of storage, with the decrease enhancing for around 0.5 log CFU/ml when compared with the probiotics which are unencapsulated.[41] Chavarri et al.[5] observed the use of chitosan as a coating material in alginate microparticles and quercetin as a prebiotics to encapsulate probiotics such as *Lactobacillus gasseri* and *B. bifidum* and they found enhanced survival in the gastrointestinal tract when exposed to the adverse condition.

Ghosh et al.[13] observed that microencapsulated probiotic *Enterobacter* sp. did not essentially weaken its defensive properties when compared with the administration of nonencapsulated *Enterobacter* sp. Ding and Shah[8] studied the viability of nonencapsulated and encapsulated probiotic organism in the presence of 3% (w/v) oxgall, 3% (w/v) taurocholic acid, pH of 2.0, and temperature of 65 C. They observed that the viability of nonencapsulated probiotic bacteria decreased more when compared to the microencapsulated probiotic bacteria.

1.5 BENEFITS OF PROBIOTICS IN AQUACULTURE

The beneficial effects of some of the probiotics in aquaculture industry are listed in Table 1.1.

1.5.1 ENHANCEMENT OF IMMUNE RESPONSE

Enhancement of the immune system of the host is an important parameter in the aquaculture. By optimizing the immune response of the aquatic animals through probiotic feed, we can prevent the multiplication of pathogen in the intestinal tract and the aquatic animal culture environment. Humoral, cellular immune responses, and expression analyses of IL-1b, TNFα,

TABLE 1.1 Beneficial Effect of Probiotics in Aquaculture.

Probiotic bacteria	Method of administration	Fish/shrimps	Beneficial effects	References
Bacillus pumilus	Unencapsulated (through feed)	Carp (Labeo rohita)	Hematological parameters	37
Lactobacillus sporogenes	Unencapsulated (through feed)	Nile tilapia (Oreochromis niloticus)	Growth performance, nutritive physiology	[11]
Rhodobacter sphaeroides, Bacillus coagulans	Unencapsulated (through feed)	Shrimp (Penaeus vannamei)	Growth performance and digestive enzyme activity	[47]
Lactobacillus plantarum	Unencapsulated (through feed)	Nile tilapia (O. niloticus)	Growth performance and immune response	[17]
Lactobacillus rhamnosus	Encapsulated (through feed)	Nile tilapia (O. niloticus)	Growth performance, intestinal morphology, and survival against streptococcal infection	[34]
Enterobacter sp.	Encapsulated (through feed and intraperitoneal injection)	Rainbow trout (Oncorhynchus mykiss)	Control bacterial cold water diseases	[13]
Enterococcus faecalis	Unencapsulated (through feed)	Javanese carp (Puntius gonionotus)	Digestive enzyme activities, immune system response, disease resistance and short-chain fatty acid production	[1]
Shewanella putrefaciens	Encapsulated (through feed)	Gilthead seabream (Sparus aurata)	Humoral innate immune parameters	[6]
Bacillus spp.	Encapsulated (through feed and water)	Whiteleg shrimp (Litopenaeus vannamei)	Enhance growth survival and water quality	[30]

and lysozyme-C were improved when fish were fed with enriched diet consisting of *Aeromonas veronii, Vibrio lentus,* and *Flavobacterium sasangense.*[49] Telli et al.[44] observed that the incorporation of probiotic *B. subtilis* at a concentration of 5×10^6 CFU·g feed-1 improved the innate immune system such as lysozyme and phagocytic activities of macrophages of Nile tilapia. Cordero et al.[7] observed that the dietary administration of probiotic *Shewanella putrefaciens* increased the cellular peroxidase and respiratory activity of gilthead seabream at high stocking density.

1.5.2 DISEASE PREVENTION

Probiotic-supplemented diets are abundantly used in aquaculture for controlling disease, mainly bacterial disease. The bacterial disease is a significant problem in aquaculture. There are many ways to prevent aquatic animals from diseases. Probiotics have become more popular for controlling bacterial disease through the production of antimicrobial substances in many developing countries. Gupta[16] observed an increase in the survival rate when the dietary supplementation of probiotics *Bacillus coagulans, Bacillus licheniformis,* and *Paenibacillus polymyxa* to the common carp fry challenged with the pathogen *A. hydrophila* significantly increased the survival rate. Newaj-Fyzul[29] observed that isolated probiotic strain *B. subtilis* from fish intestine was effective when supplemented with feed for controlling infections caused by a fish pathogen *Aeromonas* sp. in rainbow trout. Zhao et al.[50] reported the dietary administration of probiotic *Bacillus cereus* at 10^7 CFU/g significantly improved immunity and disease resistance in juvenile *Apostichopus japonicus.*

1.5.3 GROWTH ENHANCEMENT

The most standard method for the application of probiotics is through oral route as a feed supplement. Among the various benefits of probiotics, growth enhancement of the aquatic animals is necessary for increasing the production. Swain and Mohanty[42] reported that there were higher growth and nutrient utilization in all probiotic-supplemented group of catla juveniles. Marzouk et al.[25] observed that *Oreochromis niloticus* groups which were supplemented with probiotic diet showed increased growth rate than groups those supplemented with basal diet, suggesting that growth performance and utilization of feed are enhanced by the addition of probiotics to a normal diet. Garg[11] observed an increase in the growth performance of Nile tilapia, *O. niloticus* when a dietary probiotic level was increased from 0.25 to 0.75 g/100 g of diet; further increase in the probiotic level (>0.75 g/100 g) resulted in a significant ($P<0.05$) growth depression and nutrient depletion.

1.5.4   DIGESTIVE PROCESSES

When the harmful microorganisms colonize the gastrointestinal tract of animals, it will lead to the improper digestion of food, and also, add some toxins during the digestion process. A probiotic has the ability to advance the development of helpful microbes in the digestive system that will degrade the harmful microorganism. In some case, the use of antibiotics may destroy the beneficial microbes. In such cases, the use of these probiotics can help to regenerate the beneficial microflora/bacteria. Stress, antibiotics, and nondigestible fibers could change the microbial balance in the intestine. The role of probiotics in the digestive system is to enhance food digestion, and prevent the entry of pathogens, and also secreting vitamins and essential amino acids and enzymes that help in the digestion of various food fibers. Wang et al.[48] found that the dietary addition of probiotic *Rhodotorula benthica* at three different levels 10^5, 10^6, 10^7 CFU/g feed increased digestive enzyme activity such as amylase, cellulase, and alginate activity in juvenile sea cucumber *A. japonicus* as compared with control group.

1.5.5   WATER QUALITY IMPROVEMENT

Probiotics are used as bioremediation tool for controlling the accumulation of organic wastes and also for reducing proliferation of bacterial pathogen in aquaculture pond. In concentrated and semi-escalated aquaculture rehearses, high stocking densities of fish alongside extraordinary sustaining and preparation frequently prompts the disintegration of water quality and expansion of pathogens.[31] According to the report by Venkateswara,[46] probiotics have been accounted to balance microflora, control pathogen, increase the disintegration of the unwanted organic substance, improve the ecological environment by minimizing the toxic gases such as NH_3,

N_2O, H_2O_2, methane, and so forth. It also increases the nutritional level of the host, increases the population of food organism, and also improves their immunity in the culture water. Khalil et al.[22] demonstrated that the incorporation of *Yucca schidigera* probiotic bacteria into prepared Nile tilapia feeds reduced the accumulation and production of ammonia.

1.6 CONCLUSION

In the aquaculture business, utilization of probiotics as a different option for anti-infection agents is essential. From this chapter, we have concluded that microencapsulated probiotic has the ability to increase its survivability during storage and in the gastrointestinal tract from the harsh environmental condition, whereas probiotics-supplemented diet (i.e., without encapsulation) get affected by harsh environmental condition because of the absence of a physical barrier over a probiotic cell. It is fundamental to maintain the stability, viability, and cellular concentration of probiotic bacteria during storage to protect the product effectiveness for application as a feed in aquaculture. The uptake of encapsulated probiotics by aquatic animals further improves growth, immune response, water quality, digestive process, and prevents entry of pathogens. The use of probiotics increases the production.

ACKNOWLEDGMENT

We acknowledge the support of VIT University for providing the facility.

KEYWORDS

- probiotics
- probiotic-supplemented diet
- disease control
- growth enhancement
- pathogens

REFERENCES

1. Allameh, S. K.; Ringø, E.; Yusoff, F. M.; Daud, H. M.; Ideris, A. Dietary Supplement of *Enterococcus faecalis* on Digestive Enzyme Activities, Short-Chain Fatty Acid Production, Immune System Response and Disease Resistance of Javanese Carp (*Puntius gonionotus*, Bleeker 1850). *Aquaculture Nutr.* **2017**, *23*, 331–338.
2. Argüello-Guevara, W.; Molina-Poveda, C. Effect of Binder Type and Concentration on Prepared Feed Stability, Feed Ingestion and Digestibility of *Litopenaeus vannamei* Broodstock Diets. *Aquaculture Nutr.* **2013**, *19*, 515–522.
3. Ayyat, M. S.; Labib, H. M.; Mahmoud, H. K. A Probiotic Cocktail as a Growth Promoter in Nile tilapia (*Oreochromis niloticus*). *J. Appl. Aquacult.* **2014**, *26*, 208–215.
4. Balcázar, J. L.; De Blas, I.; Ruiz-Zarzuela, I.; Cunningham, D.; Vendrell, D.; Múzquiz, J. L. The Role of Probiotics in Aquaculture. *Vet. Microbiol.* **2006**, *114*, 173–186.
5. Chavarri, M.; Maranon, I.; Ares, R.; Ibanes, F. C.; Marzo, F.; Villaran Mdel, C. Microencapsulation of a Probiotic and Prebiotic in Alginate-Chitosan Capsules Improves Survival in Simulated Gastro-Intestinal Conditions. *Int. J. Food Microbiol.* **2010**, *142*, 185–189.
6. Cordero, H.; Guardiola, F. A.; Tapia-Paniagua, S. T.; Cuesta, A.; Meseguer, J.; Balebona, M. C.; Moriñigo, M. Á.; Esteban, M. Á. Modulation of Immunity and Gut Microbiota After Dietary Administration of Alginate Encapsulated *Shewanella putrefaciens* Pdp11 to Gilthead Seabream (*Sparus aurata* L.). *Fish Shellfish Immunol.* **2015**, *45*, 608–618.
7. Cordero, H.; Morcillo, P.; Meseguer, J.; Cuesta, A.; Esteban, M. Á. Effects of *Shewanella putrefaciens* on Innate Immunity and Cytokine Expression Profile Upon High Stocking Density of Gilthead Seabream Specimens. *Fish Shellfish Immunol.* **2016**, *51*, 33–40.
8. Ding, W. K.; Shah, N. P. Acid, Bile, and Heat Tolerance of Free and Microencapsulated Probiotic Bacteria. *J. Food Sci.* **2007**, *72*, 446–450.
9. Fuller, R. Probiotics in Man and Animals. *J. Appl. Microbiol.* **1989**, *66*, 365–378.
10. Ganguly, S.; Paul, I.; Mukhopadhayay, S. K. Application and Effectiveness of Immunostimulants, Probiotics, and Prebiotics in Aquaculture: A Review. *Isr. J. Aquacult./Bamidgeh* **2010**, *62*, 130–138.
11. Garg, S. K. Effect of Dietary Probiotic Mix (SPILAC) on Growth Performance and Nutritive Physiology of Nile Tilapia, *Oreochromis niloticus* (Linn.) Under Laboratory Conditions. *Int. J. Fish. Aquat. Stud.* **2015**, *3*(2), 440–446.
12. Gatesoupe, F. J. The Use of Probiotics in Aquaculture. *Aquaculture* **1999**, *180*, 147–165.
13. Ghosh, B.; Cain, K. D.; Nowak, B. F.; Bridle, A. R. Microencapsulation of a Putative Probiotic *Enterobacter* species, C6-6, to Protect Rainbow Trout, *Oncorhynchus mykiss* (Walbaum), Against Bacterial Coldwater Disease. *J. Fish Dis.* **2016**, *39*, 1–11.
14. Gibbs, B. F.; Kermasha, S.; Alli, I.; Mulligan, C. N. Encapsulation in the Food Industry: A Review. *Int. J. Food Sci. Nutr.* **1999**, *50*, 213–224.
15. Gismondo, M. R.; Drago, L.; Lombardi, A. Review of Probiotics Available to Modify Gastrointestinal Flora. *Int. J. Antimicrob. Ag.* **1999**, *12*, 287–292.
16. Gupta, A.; Gupta, P.; Dhawan, A. Dietary Supplementation of Probiotics Affects Growth, Immune Response and Disease Resistance of *Cyprinus carpio* Fry. *Fish Shellfish Immunol.* **2014**, *41*, 113–119.

17. Hamdan, A. M.; El-Sayed, A. F. M.; Mahmoud, M. M. Effects of a Novel Marine Probiotic, *Lactobacillus plantarum* AH 78, on Growth Performance and Immune Response of Nile Tilapia (*Oreochromis niloticus*). *J. Appl. Microbiol.* **2016**, *120*(4), 1061–1073.

18. Hansen, G. H.; Olafsen, J. A. Bacterial Colonization of Cod (*Gadus morhua* L.) and Halibut (*Hippoglossus hippoglossus*) Eggs in Marine Aquaculture. *Appl. Environ. Microbiol.* **1989**, *55*, 1435–1446.

19. Institute of Medicine of the National Academies. Regulating Pre- and Probiotics: A US FDA Perspective. In *Ending the War Metaphor: The Changing Agenda for Unraveling the Host-Microbe Relationship;* Vaillancourt, J., Ed.; The National Academies Press: Washington, D.C., 2006; pp 229–237.

20. Iwashita, M. K. P.; Nakandakare, I. B.; Terhune, J. S.; Wood, T.; Ranzani-Paiva, M. J. T. Dietary supplementation with *Bacillus subtilis*, *Saccharomyces cerevisiae* and *Aspergillus oryzae* Enhance Immunity and Disease Resistance Against *Aeromonas hydrophila* and *Streptococcus iniae* Infection in Juvenile Tilapia *Oreochromis niloticus*. *Fish Shellfish Immunol.* **2015**, *43*, 60–66.

21. Kesarcodi-Watson, A.; Kaspar, H.; Lategan, M. J.; Gibson, L. Probiotics in Aquaculture: The Need, Principles and Mechanisms of Action and Screening Processes. *Aquaculture* **2008**, *274*, 1–14.

22. Khalil, R. H.; Saad, T. T.; Mehana, H.; Ragab, G. Mohammed, R. A. E. A. Effect of Yucca Schidigera on Water Quality of Nile Tilapia Fingerlings. *J. Am. Sci.* **2015**, *11*(12) 83–88.

23. Lara-Flores, M.; Aguirre-Guzmán, G. The Use of Probiotic in Fish and Shrimp Aquaculture: A Review. In *Probiotics: Production, Evaluation and Uses in Animal Feed;* Guerra, N. P., Castro, L. P., Eds.; Research Signpost: Kerala, 2009; pp 75–89.

24. Martínez Cruz, P.; Ibáñez, A. L.; Monroy Hermosillo, O. A.; Ramírez Saad, H. C. Use of Probiotics in Aquaculture. *ISRN Microbiol.* **2012**, *2012*, 916845.

25. Marzouk, M. S.; Moustafa, M. M.; Mohamed, N. M. The Influence of Some Probiotics on the Growth Performance and Intestinal Microbial Flora of *O. niloticus*. In *8th International Symposium of Tilapia in Aquaculture,* Oct., Cairo, Egypt, 2008; pp 1059–1071.

26. Mortazavian, A.; Razavi, S. H.; Ehsani, M. R.; Sohrabvandi, S. Principles and Methods of Microencapsulation of Probiotic Microorganisms. *Iran. J. Biotechnol.* **2007**, *5*, 1–18.

27. Muñoz-Atienza, E.; Gómez-Sala, B.; Araújo, C.; Campanero, C.; Del Campo, R.; Hernández, P. E.; Cintas, L. M. Antimicrobial Activity, Antibiotic Susceptibility and Virulence Factors of Lactic Acid Bacteria of Aquatic Origin Intended for Use as Probiotics in Aquaculture. *BMC Microbiol.* **2013**, *13*, 15.

28. Nageswara, P. V.; Babu, D. E. Probiotics as an Alternative Therapy to Minimize or Avoid Antibiotics Use in Aquaculture. *Fish. Chim.* **2006**, *26*, 112–114.

29. Newaj-Fyzul, A.; Adesiyun, A. A.; Mutani, A.; Ramsubhag, A.; Brunt, J.; Austin, B. *Bacillus subtilis* AB1 Controls *Aeromonas* Infection in Rainbow Trout (*Oncorhynchus mykiss*, Walbaum). *J. Appl. Microbiol.* **2007**, *103*, 1699–1706.

30. Nimrat, S.; Suksawat, S.; Boonthai, T.; Vuthiphandchai, V. Potential Bacillus Probiotics Enhance Bacterial Numbers, Water Quality and Growth During Early Development of White Shrimp (*Litopenaeus vannamei*). *Vet. Microbiol.* **2012**, *159*, 443–450.

31. Padmavathi, P.; Sunitha, K.; Veeraiah, K. Efficacy of Probiotics in Improving Water Quality and Bacterial Flora in Fish Ponds. *Afr. J. Microbiol. Res.* **2012**, *6*, 7471–7478.

32. Parker, R. B. Probiotics, the Other Half of the Antibiotic Story. *Anim. Nutr. Health.* **1974**, *29*, 8.

33. Pinpimai, K.; Rodkhum, C.; Chansue, N.; Katagiri, T.; Maita, M.; Pirarat, N. The Study on the Candidate Probiotic Properties of Encapsulated Yeast, *Saccharomyces cerevisiae* JCM 7255, in Nile Tilapia (*Oreochromis niloticus*). *Res. Vet. Sci.* **2015**, *102*, 103–111.

34. Pirarat, N.; Pinpimai, K.; Rodkhum, C.; Chansue, N.; Ooi, E. L.; Katagiri, T.; Maita, M. Viability and Morphological Evaluation of Alginate-Encapsulated *Lactobacillus rhamnosus* GG Under Simulated Tilapia Gastrointestinal Conditions and its Effect on Growth Performance, Intestinal Morphology and Protection Against *Streptococcus agalactiae*. *Anim. Feed Sci. Technol.* **2015**, *207*, 93–103.

35. Poshadri, A.; Aparna, K. Microencapsulation Technology: A Review. *J. Res. Angrau.* **2010**, *38*, 86–102.

36. Preetha, R.; Jayaprakash, N. S.; Singh, I. Synechocystis MCCB 114 and 115 as Putative Probionts for *Penaeus monodon* Post-Larvae. *Dis. Aquat. Organ.* **2007**, *74*, 243.

37. Rajikkannu, M.; Natarajan, N.; Santhanam, P.; Deivasigamani, B.; Ilamathi, J.; Janani, S. Effect of Probiotics on the Haematological Parameters of Indian Major Carp (*Labeo rohita*). *Int. J. Fish. Aquat. Stud.* **2015**, *2*(5), 105–109.

38. Salminen, S.; Ouwehand, A.; Benno, Y.; Lee, Y. K. Probiotics: How Should they be Defined?. *Trends. Food Sci. Technol.* **1999**, *10*, 107–110.

39. Senok, A. C.; Ismaeel, A. Y.; Botta, G. A. Probiotics: Facts and Myths. *Clin. Microbiol. Infect.* **2005**, *11*, 958–966.

40. Soccol, C. R.; Vandenberghe, L. P. D. S.; Spier, M. R.; Medeiros, A. B. P.; Yamaguishi, C. T.; Lindner, J. D. D.; Thomaz-Soccol, V. The Potential of Probiotics: A Review. *Food Technol. Biotechnol.* **2010**, *48*, 413–434.

41. Sultana, K.; Godward, G.; Reynolds, N.; Arumugaswamy, R.; Peiris, P.; Kailasapathy, K. Encapsulation of Probiotic Bacteria with Alginate–Starch and Evaluation of Survival in Simulated Gastrointestinal Conditions and in Yoghurt. *Int. J. Food Microbiol.* **2000**, *62*, 47–55.

42. Swain, S. K.; Mohanty, S. N.; Tripathi, S. D. Growth and Survival of Catla Spawn Fed a Dry Artificial Diet in Relation to Various Stocking Densities. In *Proceedings of National Symposium on Aquacrops,* Central Institute of Fisheries Education: Bombay, 1994, 55.

43. Tapia-Paniagua, S. T.; Díaz-Rosales, P.; León-Rubio, J. M.; de La Banda, I. G.; Lobo, C.; Alarcón, F. J.; Chabrillón, M.; Rosas-Ledesma, P.; Varela, J. L.; Ruiz-Jarabo, I.; Arijo, S. Use of the Probiotic *Shewanella putrefaciens* Pdp11 on the Culture of Senegalese sole (*Solea senegalensis*, Kaup 1858) and Gilthead Seabream (*Sparus aurata L.*). *Aquacult. Int.* **2012**, *20*, 1025–1039.

44. Telli, G. S.; Ranzani-Paiva, M. J. T.; de Carla Dias, D.; Sussel, F. R.; Ishikawa, C. M.; Tachibana, L. Dietary Administration of *Bacillus subtilis* on Hematology and Non-Specific Immunity of Nile Tilapia *Oreochromis niloticus* Raised at Different Stocking Densities. *Fish Shellfish Immunol.* **2014**, *39*, 305–311.

45. US Food and Drug Administration. *Direct-Fed Microbial Products;* Section 698.100 (Revised March 1995), 2007.

46. Venkateswara, A. R. Bioremediation to Restore the Health of Aquaculture. *Pond Ecosystem Hyderabad* **2007**, *500*, 1–12.
47. Wang, Y. B. Effect of Probiotics on Growth Performance and Digestive Enzyme Activity of the Shrimp *Penaeus vannamei*. *Aquaculture* **2007**, *269*, 259–264.
48. Wang, J. H.; Zhao, L. Q.; Liu, J. F.; Wang, H.; Xiao, S. Effect of Potential Probiotic *Rhodotorula benthica* D30 on the Growth Performance, Digestive Enzyme Activity and Immunity in Juvenile Sea Cucumber *Apostichopus japonicus*. *Fish Shellfish Immunol.* **2015**, *43*, 330–336.
49. Xu, Y. J.; Wang, Y. B.; Lin, J. D. Use of *Bacillus coagulans* as a Dietary Probiotic for the Common Carp, *Cyprinus carpio*. *J. World Aquacult. Soc.* **2014**, *45*, 403–411.
50. Zhao, Y.; Yuan, L.; Wan, J.; Sun, Z.; Wang, Y.; Sun, H. Effects of Potential Probiotic *Bacillus cereus* EN25 on Growth, Immunity and Disease Resistance of Juvenile Sea Cucumber *Apostichopus japonicus*. *Fish Shellfish Immunol.* **2016**, *49*, 237–242.

CHAPTER 2

EXPERIMENTAL INVESTIGATION OF CARBONIZATION PROCESSES IN THE FORMATION OF FOAM-BASED EPOXY RESIN CONTAINING CARBON NANOPARTICLES

S. G. SHUKLIN[1,*], D. S. SHUKLIN[1], and A. V. VAKHRUSHEV[1,2]

[1]*Kalashnikov Izhevsk State Technical University, Izhevsk, Russia*

[2]*Institute of Mechanics, Ural Division, Russian Academy of Sciences, Izhevsk, Russia*

Corresponding author. E-mail: shuklin_sg@mail.ru

ABSTRACT

The directions of the creation of heat-protective polymeric materials are described in the work. The possibility of predicting the properties of fireproof polymer materials is shown. The scientific principles of reducing the combustibility of polymer materials using carbon nanostructures are substantiated. The comparative estimation of variants of fire protection of polymeric materials is given. The efficiency of using nanostructures in polymer materials and their use in intumescent compositions are analyzed. The influence of carbon nanostructures on the property of foam boxes was studied. The thermophysical characteristics of the foam boxes have been determined and it has been proved that the use of carbon nanostructures makes it possible to significantly increase the strength characteristics and the multiplicity of foaming of the foam boxes. The foam boxes were investigated by such methods as atomic force microscopy, X-ray photoelectron spectroscopy, X-ray fluorescence. It is proved that carbon nanostructures are active structure-forming agents. It is determined that carbonization begins on the inner surface of the bubble when modified by carbon nanostructures. The process of carbonization of the polymer material is significantly accelerated when exposed to fire and heat sources.

2.1 INTRODUCTION

Following the analysis of literature review, there are several problems in creating heat-protective composite polymer materials (CPMs).

1. Insufficient development of experimental and mathematical modeling of burning processes of polymeric materials, in particular, flame retardant CPM, in some measure is caused by the complexity of the processes themselves and the features of CPMs.
2. Basically, these are means for determining chemical reactions on the surface and in the interphase layers of flame retardant CPM, without which it is difficult to obtain a fire protection mechanism.
3. Insufficient development of research on the technology of processing fire retardant materials and the interaction of material components among themselves, leading to changes in the properties of materials, including the characteristics of fire safety.

In some cases, modification by fire retardants or fire retardant systems can lead to deterioration of the basic characteristics of materials. There are many reasons of the change in characteristics at introducing fire retardants. They can be grouped according to the following characteristics:

1. Antagonism of the additives and matrix, leading to phase separation during storage, operation, and processing
2. The antagonism between the additives introduced during the processing or obtaining the material leads to a decrease in the certain positive effects of some or most of the additives
3. The antagonism of the additives and "reinforcing," including reinforcing filler, leads to a decrease in the adhesive interaction of the polymer matrix and filler, especially during the operation
4. The influence of harmful impurities and the passivation of additives when exposed to the external environment and the processes of storage, transportation, and joint processing, leads to antagonism of the first, second, or third type.

Moreover, it follows from the analysis of the literature data that insufficient attention is paid to the investigation of processes occurring during the expansion of CPM. The changes in the chemical structure of the coating, the effect on the process of foam, and the coke formation of various components, as well as the heating conditions have not been studied to the end. A limited amount of works devoted to these problems is due to the complexity of the processes under consideration, the lack or high cost of the necessary technical means.

One of the ways to study the processes taking place in the material is mathematical modeling and carrying out a computational experiment. The description of the mathematical model of physicochemical processes in intumescent polymer coatings is also difficult due to the complexity of the processes themselves. Earlier, attempts were made to perform mathematical modeling, but the chemical reactions taking place in the coating were not considered, and parameters such as temperature and rate of increase in the height of the expanded layer were estimated approximately on the basis of experimental studies.

Recent papers show the effect of additives of carbon nanostructures, in particular, nanotubes (NTs) and fullerenes on the structure, mechanical, and flame retardant properties of polymers.[1] It is found that carbon nanostructures can serve as an effective structuring additive that increases the physical and mechanical characteristics of polymer composites.[2] Fullerenes C60, when added to polystyrene, act as a plasticizing additive, reducing the mechanical stresses arising during deformation.[3,4] However, the use of carbon nanostructures is constrained by their high cost. In addition, relatively little data has been accumulated on this issue so far, which do not allow one to fully appreciate the effectiveness of polymer modification by carbon nanostructures. The carbon NTs (CNTs) combine the properties of molecules and a solid and can be regarded as an intermediate state of matter. This feature attracts the constant attention of researchers studying the fundamental features of the behavior of such an exotic object under different conditions. These features, which are of considerable scientific interest, can be used as the basis for the effective applied use of NTs in various fields of science and technology.

The high mechanical strength of CNTs in combination with their electrical conductivity makes it possible to use them as a probe in a scanning microscope designed to study the smallest surface inhomogeneities.[5] This increases the resolution of these types of instruments by several orders of magnitude and put them on a par with such a unique device such as field ion microscope. The use of NTs in chemical technology has significant prospects. One possible approach of this kind, based on the high specific surface area and chemical inertness of CNTs, is associated with the use of NTs in the heterogeneous catalysis as a substrate. The results of preliminary experiments[6] indicate anomalously high catalytic activity of NTs during liquid-phase hydrogenation of nitrotoluene and gas-phase hydrogenation of CO in the presence of nickel and praseodymium nanoparticles embedded in NTs material.

Currently, epoxy resins are widely used for the manufacturing of structural and functional composite materials, adhesives, and sealants. In connection with this, the task of increasing their physical, mechanical, and operational characteristics is extremely urgent. One of the directions within which the task can be solved is the modification of epoxy matrices by nanoparticles. CNTs, due to their high mechanical characteristics, as well as the wide possibilities of surface functionalization,[7] providing covalent interaction with the polymer matrix, can be considered as one of the most promising types of modifiers.

Owing to the high specific surface, the increase in the mechanical properties of composites is achieved already at low concentrations of CNTs while improving the thermal and electrically conductive properties of the material. This explains the exceptional interest shown by the researchers in questions related to the methods of introducing CNTs into epoxy resins, with the development of options for the functionalization of the surface of CNTs, and the search for modification conditions that provide a directional change in the properties of the polymer matrix.

At present, it is considered that the role of CNTs in improving the physicomechanical characteristics of epoxy nanocomposites is to effectively redistribute the load in the volume of the binder. In this case, the main conditions for the efficiency of modification are:[8]

- Uniform distribution of CNTs by the volume of the matrix
- Ensuring high adhesion of the CNT surface to the polymer matrix
- Orientation of CNTs in the direction of load action

However, a number of experimental data suggest that reinforcement of the polymer matrix of CNTs is not the only reason for improving the physical and mechanical properties of epoxynanocomposites. The influence of CNT on the kinetics of curing of epoxy oligomers and the structure of the resulting composites is also significant.

Thus, it has been established[9] that isothermal heating conditions do not allow the complete composition of tetraglycidyl-4,4-diaminodiphenylmethane to be completely cured in the presence of 4,4-diaminodiphenylsulfone as a curing agent at a curing temperature of 190°C, the degree of conversion of the composition modified with 5% CNT, is 10% higher than the original system. However, at a curing temperature of 220°C, the situation becomes opposite: the degree of conversion of the initial system is 4% higher than the degree of conversion of the modified composition.[10]

There are also mentions of the modification of unsaturated polyester resins, where CNTs lead not only to a change in the physicomechanical parameters of the composite but also accelerates the process of gelling.[11,12]

A huge number of studies on the chemical and physical modification of NTs have been and continue to be carried out with the sole purpose of achieving a better alignment and uniform distribution of NTs in the polymer matrix. These works yield their results, such as modification of NTs by fluorine,[13] but the uniform particle distribution in the volume is solved only with the use of special equipment and/or ultrasound. This method is limited by the volume of the modified material. It is long enough and requires verification of the result after each processing of a small batch, if not to recognize the customized technology of particle distribution.

An important factor affecting the combustibility of polymer materials is the physical structure of the foam box. To solve this problem, it is proposed to use in the intumescent coatings, as carbon foam structuring agents in small quantities, carbon–metal-containing nanostructures that would allow the structure of the foam coke to be ordered and the physicochemical characteristics of the foam coke and CPM to be increased. When setting the problem, it is important to take into account the changes in the thermophysical characteristics of polymer nanocomposites and foam boxes. The processes of foam and coke formation in intumescent coatings are investigated. The paper presents a comprehensive approach to the problem of fireproofing of polymeric materials, as a result of this approach it becomes possible to predict the properties of fireproof polymer materials.

2.2 SCIENTIFIC BASIS FOR THE PRINCIPLES OF REDUCTION FLAMMABILITY OF POLYMER MATERIALS USING NANOSTRUCTURES

In accordance with the accumulated experience, it can be concluded with certainty that the main options for the fire protection of combustible materials are to reduce the probability of an exothermic redox reaction, which can be expressed by reducing the concentrations of the oxidant and/ or reducing agent below a certain limit, by lowering the temperature below the critical temperature. This implies significant changes in the conditions of ignition and combustion.

2.2.1 COMPARATIVE EVALUATION OF FIRE PROTECTION OPTIONS POLYMERIC MATERIALS

The methods for protecting polymeric materials from high-temperature flows or flames are known[14] by using flame retardant coatings or by introducing fire retardants into materials. However, both of these ways of reducing flammability have certain drawbacks. In the first case, it is possible to peel off the coating during operation or the source of combustion. In the second case, a change (sometimes significant) in the main characteristics of the material sharply reduces the area of its application. Therefore, when using flame retardant coatings, the change in thermal and adhesive characteristics during the fire action plays a big role, and when fire retardants are introduced into polymeric materials, it is necessary to reduce the content of additives added to the material to an effective minimum, thereby increasing their activity in reducing flammability.

In either case, nanophases to a certain extent stimulate the formation of organic or inorganic cokes that prevent the destruction of the bulk of the polymer material. Combustion of flame-retardant polymeric materials has a mixed combustion pattern,[15] that is, when coking, regularities of heterogeneous combustion are manifested, when unstable combustion is noted, provided that the coked surface will occupy more than 80% of the surface subject to combustion.[16] When active additives, which are nano-structures that can exhibit the properties of nanocatalysts, are introduced into polymer matrices, one can expect the formation of nanophases that stimulate the self-organization of macromolecules of the matrix into the "preforms" of the future coked surface (S_c). Then S_c can be written in the form of a formula reminiscent of the Silberberg equation[17]:

$$S_c = LEk_s k_e \qquad (2.1)$$

where L is the coordination number of the active element of the fire retardant involved in the interaction or the maximum possible number of active groups in the retardant or in the fire-retardant system; E is the optimum content of the active element in the retardant; K_s is the number of macromolecules of the first layer adsorbed on the active surface of the flame retardant; K_e is the number of macromolecules of the second layer entering the nanophase formed.

The product depends on the adsorption options and the interactions between the flame retardants and macromolecules of the matrix polymers. The reduction of the particles of the fire retardant and uniform distribution of them in the material promotes an increase in the nanophase and "preparation" of the coked surface and, correspondingly, a decrease in the content of the active element. As an example, the values for combustibility of phosphorus-containing polyesters and semiempirical calculations are given, which follow from the following positions:

1. An increase in the content of active groups or an active element increases their intermolecular interaction, which leads to the association and formation of aggregates of particles whose surface is depleted by actively interacting centers with macromolecules.
2. Reduction of the content of the active element with an increase in the degree of dispersion, in contrast to the previous one, enhances the chemisorption and structuring of polymer macromolecules by analogy with structure-forming agents.[18]

Since the coked surface cannot exceed 100%, the number of active phosphoric groups does not exceed more than four, and the experimental effective phosphorus content is determined to be 5.9%,[14] when the magnitude of the product $k_s k_e$ is close to four. At the same time, as the number of adsorbed molecules on active sites decreases, the number of changes in the structure of the second layer of macromolecules increases, that is, the coefficient K_s, characterizing the chemisorption of macromolecules on the active surfaces of the flame retardant should decrease with the content of the active element, and the coefficient K_e, characterizing the degree of aggregation, increase.

From these provisions, it follows that the dependences of flammability and other characteristics, for example, physical and mechanical properties and density on the content of the fire retardant, will be described by curves with extrema (minima and maxima). The flammability rating of the loss of mass during combustion can be inversely proportional to the area (%) of the coked surface in the combustion zone (Fig. 2.1).

The solid curve in Figure 2.1 is calculated on the basis of the given positions, and the dashed curve is the experimental data for phosphorus-containing polyesters of a closed structure (Table 2.1) at $S_e \sim (100-m)\%$.

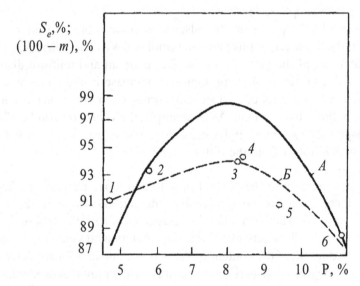

FIGURE 2.1   Dependence of S_e on the content of phosphorus.

TABLE 2.1   Change in Mass Loss During Combustion with Change Phosphorus Content.

Phosphorus-containing polyesters	Content phosphorus (%)	Mass loss during combustion, m (%)	100−m (%)
PPE 1	4.7	10	90
PPE 2	5.7	6.5	93.5
PPE 3	8.0	5.5	94.5
PPE 4	8.2	5.4	93
PPE 5	9.0	7.0	93
PPE 6	11.2	11.5	88.5

Of course, there is no complete coincidence of the curves, but their form and position of the maxima are similar.

The same dependences were obtained experimentally for phosphorus–vanadium-containing polyamides (Fig. 2.2). In this case, the combustibility is determined by the time of independent combustion, and the physicomechanical properties are represented by the destructive stress in bending.

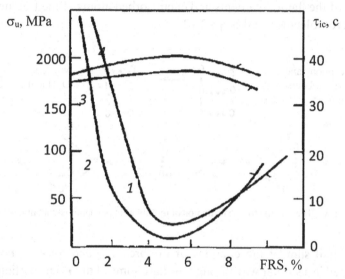

FIGURE 2.2 The change in the self-combustion of Vcd (1, 2) and the destructive (3, 4) depending on the content in the polyamides (kapron 1, 3 and nylon 2, 4) of the PGS [P-Ba (NO$_3$) 2].

The content of the fire retardant and the ratio of components in complex fire-retardant systems are proposed to be estimated by taking into account the space-energy parameters. If a protective carbon layer is formed on the surface of the material or coating, then the applied fire retardants must be transformed into a complex structure with an interlayer distance close to or slightly larger than the interlayer distances of graphite-like substances.

Energy characteristics must correspond to the transition of chemical bonds of one kind to another. In this case, if the half-sum of the collapsing and forming bonds is less than the adsorption potential or is equal to it, then the process is possible. This approach is based on the concepts of the multiplet theory of catalysis. For polyesters containing phosphoric acid groups, one can write the schemes of bond transformation in polyester with the participation of P–O pairs (Fig. 2.3).

The interlayer distances d (polyphosphates or polyphosphoric acid) are taken to be equal to the sum of the van der Waals radii $d = Rp + R0 = 3.3$ nm. The adsorption potential q is expressed in terms of the half-sums of the electronegativities of the atoms participating in the interaction of the active centers of the flame retardants and chemisorbed atoms of the fragments of the macromolecules, and is $q = 2.72$.

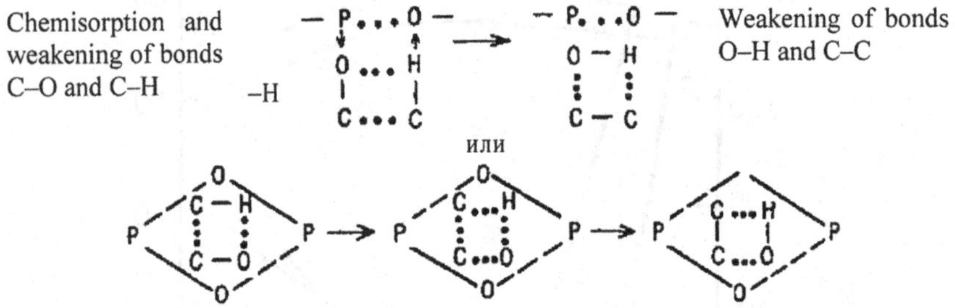

FIGURE 2.3   The scheme for the formation of multiplets in phosphorus-containing polyester.

The half-sum of the energies of the breaking and forming bonds is calculated in a similar way through the half-sums of the atoms participating in the bonds. For polyesters, $\sum E/2 \approx 2.65$. Since $q \geq \sum E/2$, the reaction is directed to the preferential formation of water and carbon substances. The use of phosphorus–vanadium oxide fire-retardant systems in evaluating carbonization products prove to be more advantageous in comparison with phosphoroxidic systems or vanadium oxide systems since the percentage of the retardant phosphorus–vanadium-containing system is required less because of the increase in the coordination number L. Depending on the coordination saturation of vanadium, the content of the coked surface changes. The thickness of the boundary layers grow and the germ "zak" of the surface increases. Thus, consideration of the coordination processes in the formation of compositions and fire-retardant systems makes it possible to predict and evaluate the effectiveness of the fireproof material or coating. In this case, the evaluation is carried out at the level of molecular and atomic interactions with the formation of nanophases, the occurrence of which is due to the formation of boundary layers.

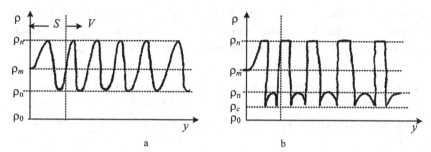

FIGURE 2.4   The density diagram with the formation of nanophase high-density (a) and low-density nanophases (b).

2.2.2   EFFECTIVENESS OF NANOSTRUCTURES IN POLYMERIC MATERIALS

The formation of nanophases and nanostructures of various composition and structure is possible with the introduction of fire-retardant systems in compositions in the form of powders or liquids. In the case of displacement of liquid components, their thermodynamic compatibility with the formation of extended phases is possible. However, it is known[19] that even polymer homologists are bound to dissolve into each other. Therefore, in multicomponent composites, the isotropy and density of materials are ensured by the distribution of components and phases from the surface to the volume of the samples. In this case, the boundary layers that are formed have properties slightly different from those of the contacting phases and which can be identified with nanophases. The density of nanophases may be intermediate between the densities of the interacting phases (Fig. 2.4a) or below the density of the one and the other phase, interactions between which do not occur (Fig. 2.4b).

As the sizes of the component phases decrease, their activity and interaction increase, which leads to an increase in dense nanophases. If the dimensions of the phases of the components are within the limits of nanophases, then the density of the composite as a whole increase along with the increase in the heat capacity due to the appearance of dense boundary layers (nanophases). Therefore, the formation of nanostructures in composites can lead to a decrease in the thermal conductivity of the material. This is confirmed by experimental data on the thermophysical characteristics of the elastic material, in which nanophases are formed upon introduction of the retardant phosphorus–vanadium-containing system, and with increasing temperature from 20 to 200°C the heat capacity changes more than the thermal conductivity (Table 2.2).

TABLE 2.2 Change in the Heat Capacity and Thermal Conductivity of the Initial Component and the Material-containing Nanostructures with a Change in Temperature.

Thermophysical characteristics	Not modified material	Modified material
C_p, kJ/kg · grad	1.47→1.89	1.66→2.54
	30→200°C	20→200°C
λ, W/m · grad	0.26→0.39	0.27→0.33
	30→200°C	30→200°C

In the case of the introduction of fire-retardant layered structures or forming a layered structure during distribution in a polymer matrix, conditions may arise for the formation of extended nanophases in which one of the dimensions extends beyond the limits of nanometer formations. These can be fibrillar, lamellar, or tubular structures having a length of 1 μm or more. In such formations as nanoreactors, it is possible to transform chemical particles with the formation of new nanophases, which increase the overall density and stability of materials to the effects of thermal and fire sources.

Moreover, as follows from the literature data, for example,[20] there are chains of interactions, and a secondary one is formed on the primary structure—"induced." Formation in the composition with the introduction of flame retardants of nanophases that promote the heterogeneous catalysis of coke formation leads in a number of cases to their "memorization" and the repetition of the corresponding reflections in the diffractograms of pyrolysis or coke residues (Fig. 2.5). As can be seen from the figure, the wide-angle diffractograms of the pyrolysis residues of flame-retardant polycarbonates retain the nanophase, which appeared when a phosphorus-containing fire-retardant system was introduced.

Thus, the nanophases in polycarbonate are resistant to high thermal loads. Therefore, it can be expected that the formation of nanocomposites containing a large number of nanophases can lead to an overall increase in their stability and to mechanical and thermal loads.

At the same time, the presence of directed processes along the formed layers in the material can lead to reinforcement of the material and the appearance of anisotropy. For fire-retardant coatings, this property contributes to a heat sink in a certain direction over the surface. The appearance in a material of nanostructures of a certain orientation contributes to the formation of double electrical layers and catalytic processes for the formation of flame-retardant coatings. Promising are fire-protective intumescent coatings, the introduction of nanostructures into which it is possible to increase the adhesion of the coating to the protected polymeric material and the strength of the formed foam coke.

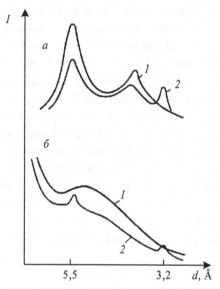

FIGURE 2.5 Wide-angle diffractograms of polycarbonates (a) and corresponding pyrolysis residues (b): 1—not containing; 2—containing the phosphorus–vanadium-containing system.

What is the mechanism of action of the nanostructures formed or introduced in this case? When flame-retardant polymer materials are formed by introducing flame retardants of combustion into polymeric compositions, which are structure-forming agents and stimulators of formation in nanophase materials, an increase in the "embryos" of the coked surface is observed with increasing temperature. In turn, the rapid creation of a coked surface layer, which has good adhesion to the bulk of the material, hinders the development of the combustion process when one and the same fire source is exposed. The most often increases the heat sink over the surface, which in a number of cases leads to a substantial decrease in the mass burning rate with a relatively high flame propagation velocity. In these cases, it is necessary to conduct additional treatment of polymer surfaces with thermal shocks. Moreover, in the interlayer spaces, original nanoreactors, additional nanostructures appear that contribute to reducing the amount of readily volatile and flammable products released when materials are exposed to fire sources.

The density of the surface layers and the heat capacity increase significantly, and the thermal conductivity along the normal to the surface decreases, therefore a sharp decrease in the thermal diffusivity is observed.

The introduction or formation of nanostructures in fire-retardant intumescent coatings results in the formation of foam boxes with regularly located closed pores and sufficiently strong bubble walls. Moreover, depending on the location and shape of the nanostructures, the gas bubbles from the gassing agent are distributed according to the mass of the foam coke in a certain order. If fine dispersed powders of ammonium polyphosphate (PFA) or its derivatives are used as the gasifier and dispersed by ultrasonic field in the composition, it is possible to approach the model system in which the distribution of the particles of the retarder and the blowing agent, such as PFA, is close to uniform (Fig. 2.6). In this figure, a simplified arrangement of particles is presented. We assume the shape of the particles in the form of balls, the size of which approximately corresponds to 10 μm, and the distance between them is close to 1 mm.

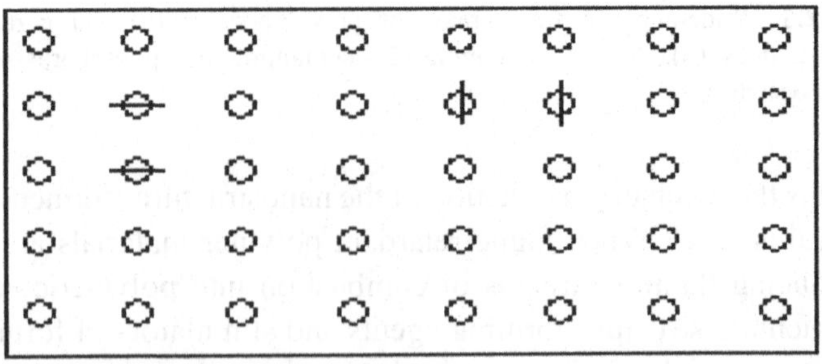

FIGURE 2.6   The PFA distribution of the epoxy composition (EC).

The formed foam box is a carbon material with bubbles filled with a mixture of ammonia and water. In this case, between the bubbles, a carbon barrier is formed, containing a layer of polyphosphoric acid on the inner surface. Unlike previous coatings in which, according to X-ray photoelectron spectroscopy (XPS) data, carburization of the surface takes place, in the intumescent coatings, carbonization of the inner wall of the resulting bubbles takes place previously.[16] Introduction to intumescent flame-retardant compositions of carbon-containing nanostructures obtained from polyvinyl alcohol (PVA) and metal salts,[17,18] increases the regularity of the location of the PFA-generated bubbles in the foam boxes. At the same time, the strength of the formed foam box increases. This result is due to a decrease in the rate of formation of ammonia bubbles due to the coordination of ammonia molecules on metal clusters, for example, copper, nickel, or cobalt, which are located in carbon shells open at the ends. Such carbon-containing metal tubulins are filled with metal phases by 60–70%. They are mostly interlaced branched structures with a diameter of 40–60 nm and a length of 300–1000 nm. The resulting bubbles from PFA are partially sorbed on the surface of these structures. At the same time, the foam cells in the process of their formation become more resistant to mechanical stresses due to the formation on the "frame" of the introduced nanostructures of new carbon nanophases. Thus, with a slight decrease in the degree of expansion, the strength of the foam box and its stability at various loads increases.

2.3 STUDY OF THE EFFECT OF CARBON-CONTAINING METAL NANOSTRUCTURES ON THE PROPERTIES OF FOAM BOXES

The idea of protecting the material from fire by forming a coke "cap" on its surface was brought to its logical conclusion when so-called swelling coatings began to be developed and applied.[21]

2.3.1 STUDY OF THE EFFECT OF CARBON-CONTAINING METAL NANOSTRUCTURES BASED ON POLYVINYL ALCOHOL GELS ON THE PROPERTIES OF FOAM BOXES

On the basis of literature data and experiments, foam coke with high flameproof properties should have uniformly distributed fine pores, the walls of which have high strength characteristics.[20,22,23] Moreover, good adhesion of the intumescent coating to the substrate or to the article to be protected is necessary.

To control the process of forming a foam box it is necessary:

1. Control the rate of destruction of the blowing agent
2. Control the uniform distribution of gas bubbles and their size in a polymer matrix
3. Manage the structure of the foam box
4. Control the processes of carbonization, which affect the strength characteristics of the bubble wall.

Nanoscale structures with a developed surface have high surface energy. Thus, the surface area of multilayered carbon NTs (MNTs) obtained by the CVD method at 700–900°C is ~400 m²/g.[24] The fraction of intergranular interfaces in the total volume of the particle[25] is calculated by the formula

$$\frac{\Delta V}{V} \approx \frac{6\delta}{d}, \tag{2.2}$$

where δ is the thickness of the interface; d is the diameter.

For particles with $d = 10$–100 nm, the fraction of the surface volume is 50–10%. The effect of small additives (up to 5 mass%) of carbon-containing metal nanostructures obtained by the above method carbonization of gels containing copper-II (II), nickel (II) and cobalt (II) aluminum oxide (IIA) copper oxides, on the expansion of modified epoxy compositions (EC), as well as the physicomechanical and thermophysical properties of the resulting carbon foams (foam boxes). The starting epoxydiene resin was modified with ammonium polyphosphate to give it flame-retardant properties. The products of carbonization of PVA gels processed at 450°C for 2 h were added to the resulting mixtures. Foam cells obtained from a modified resin without additive and adding carbonization products of PVA gels, ECs with nanostructures (ECNS) were studied (Table 2.3).

It was originally assumed that nanostructures can serve (due to nanoscale and wall consisting of a layered carbon phase) as effective carbon crystallization centers formed during carbonization of ECs (due to the "spider-like" structure), effective structure-forming agents that enhance the mechanical properties of foam boxes.

TABLE 2.3 Compositions.

Component	Number composition											
	1	2	3	4	5	6	7	8	9	10	11	12
ED-20	10.0	10.0	10.0	10.0	10.0	10.0	10.0	10.0	10.0	10.0	10.0	10.0
PFA	3.0	3.0	3.0	3.0	3.0	3.0	3.0	3.0	3.0	3.0	3.0	3.0
Gel No. 1	0.5											
Gel No. 3		0.5										
Gel No. 5			0.5									
Gel No. 6				0.5								
Gel No. 8					0.5							
Gel No. 10						0.5						
Gel No. 11							0.5					
Gel No. 13								0.5				
Gel No. 15									0.5			
graphite										0.5		
C_{act}*											0.5	
PP**	1.5	1.5	1.5	1.5	1.5	1.5	1.5	1.5	1.5	1.5	1.5	1.5

*Activated carbon.

**Polyethylene polyamine.

2.3.1.1 INVESTIGATION OF THE PROPERTIES OF FOAM BOXES

In the temperature range up to 500°C, a peak of mass loss (~300°C) is observed on the derivatograms of the compositions, the shape of the THM curves for samples containing Ni and Co is similar; for samples containing additives with the same metal, the shape of the THM curves is also similar (Fig. 2.7).

In the process of continuous heating, the heat capacities of the compositions (before the expansion) and the foam boxes (after the start of expansion) were measured at temperatures of 25–400°C in 25°C increments on an IT-S-400 instrument. A smooth line without markers in Figures 2.8 and 2.9 is obtained by approximating the Cp (T) dependence by a fifth-degree polynomial by the method of least squares. Curves, similar, to those are shown in Figures 2.8 and 2.9, are observed for all epoxidian compositions containing a single-metal additive.

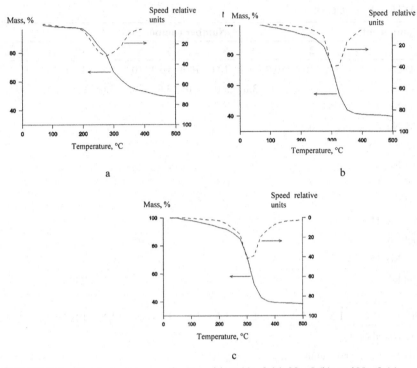

FIGURE 2.7   The derivatograms of compositions No. 2 (a), No. 5 (b), and No. 8 (c).

As can be seen from Figure 2.2 and 2.3, the curves of Cp (T) pass through a maximum, and for compositions containing nanostructures with Cu, Ni, and Co, they have a somewhat different form. For copper foams with a maximum, the maximum value of Cp does not exceed ∼2000 J/K, while for foam-coxes with nickel and cobalt it reaches ∼4000–4500 J/K. The heat capacity of the foam boxes from ECs without additives reaches ∼6000 J/K. As can be seen from Figure 2.9, the maximum of Cp (T) for ECs without additives corresponds to a temperature of ∼170°C, whereas for samples of ECs with nanostructures Nos. 2, 5, and 8 is in the range 250–280°C (Fig. 2.8a and c). That corresponds to the beginning of the mass loss interval on the TGM curves (Fig. 2.7). The growth of the heat capacity in these temperature ranges can be explained by the presence of gas bubbles in the samples formed during carbonization of epoxy resin in the presence of ammonium polyphosphate. According to[26] the following, transformations are typical for ammonium polyphosphate upon heating:

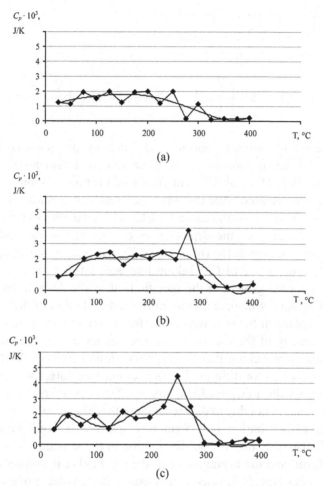

FIGURE 2.8 Dependences of heat capacity Cp on temperature for compositions No. 2 (a), No. 5 (b), and No. 8 (c).

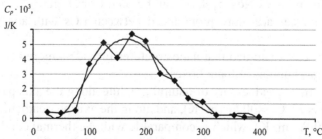

FIGURE 2.9 Dependence of heat capacity Cp on temperature for composition No. 12.

$$\begin{bmatrix} O \\ \parallel \\ -P-O- \\ \mid \\ ONH_4 \end{bmatrix}_n \xrightarrow{t^o} \begin{bmatrix} O \\ \parallel \\ -P-O- \\ \mid \\ OH \end{bmatrix}_n + nNH_3$$

$$R-CH_2-CH_2-OH + \begin{bmatrix} O \\ \parallel \\ -P-O- \\ \mid \\ OH \end{bmatrix}_n \xrightarrow{t^o} R-CH=CH_2 + H_3PO_4 + \begin{bmatrix} O \\ \parallel \\ -P-O- \\ \mid \\ OH \end{bmatrix}_{n-1}$$

The resulting orthophosphoric acid, such as polyphosphoric acid, promotes the dehydration of hydroxyl-containing fragments of organic macromolecules. Thus, at different stages of heating, gaseous ammonia and water are released. Accordingly, two peaks of the maximum heat capacity are observed on the curve for ECs without additives (Fig. 2.9).

It can be seen from the dependences of Cp (T) that the addition of nanostructures significantly (by ~80–100°C) increases the temperature of the maximum value of Cp. Taking into account that the gas phase is responsible for the growth of the specific heat, it should be assumed that when adding nanostructures a more uniform distribution of the produced gases takes place in the structure of the foam boxes, as a result of which the heat capacity of the foam boxes decreases relative to the unblended foam boxes, approaching that for the solid (initial ECNS). At the same time, the mass loss for different ECNs is close; therefore, the decrease in the maximum value of Cp for foam box No. 2 in comparison with the Nos. 5, 8 foam boxes is not due to different amounts of the gases formed, but to their different distribution in the foam box structure. It should be assumed that for the No. 2 foam, the gas distribution in the structure is somewhat more uniform, and the average size of the gas bubble is smaller than for the foam boxes Nos. 5, 8, since in this case, a large volume of the bubble surface in the foam box is due to a certain volume of gas, which leads to lower, average in volume, heat capacity of the foam box in comparison with the foam boxes Nos. 5, 8. As can be seen from Figures 2.8 and 2.9, these differences are more pronounced between ECs with and without additives.

Thus, it can be argued that nanostructures provide a more even distribution of gas bubbles in foam boxes.

The thermal conductivity of natural graphite along the axis parallel to the layers is 25°C ~400 W/(m · K), and along the axis perpendicular to the layers ~80 W/(m · K), which is comparable with the thermal conductivity of copper (388 W/(m · K)).[27] The thermal conductivity of polycrystalline graphite usually does not exceed ~200 W/(m · K), but can reach ~500 W/ (m · K). The thermal conductivity of the resulting foam boxes is approximately the same and the thermal conductivity of polycrystalline graphite is ~550 times smaller on average (Table 2.2). The higher the heat-shielding properties of the foam boxes, the lower is its thermal conductivity. In this respect, the most effective are foam boxes of systems Nos. 4–6, 9 having the minimum values of the indicated value. It should be noted that the theoretically calculated thermal conductivity of isolated ONT symmetry (10, 10) is very high and is ~6600 W/(m · K), but the value obtained for compact samples is much lower because of the interlayer interaction and does not exceed ~0.7 W/(m · K),[28] therefore the contribution of small nanostructured additives to the increase in thermal conductivity is very small. The specific heat capacities of the foam boxes (Table 2.4) are an order of magnitude higher than the heat capacity of graphite at room temperature (~720 J/[kg · K]),[27] which is due to the porous structure of the foam boxes and the significant contribution of the heat capacity of the gas in the pores.

TABLE 2.4 Thermal Conductivity (λ) and Specific Heat Capacity (Cm) of the Foam Boxes Obtained from Compositions Nos. 1–12 at 800°C.

Number composition	C_m, J/(kg·K)	λ, W/(m·K)	No. п/п	C_m, J/(kg·K)	λ, W/(m·K)
1	6146	0.384	7	5763	0.385
2	6216	0.382	8	5773	0.363
3	4916	0.412	9	7470	0.350
4	5481	0.332	10	5975	0.384
5	5772	0.362	11	5784	0.388
6	5881	0.329	12	5176	0.377

The true densities of foam boxes with nanostructures do not exceed ~1.4 g/cm³ and are practically independent of the type of metal that is part of the nanostructures (Table 2.5).

The maximum possible density of graphite is 2.265 g/cm³; nevertheless, the density of even natural graphite rarely exceeds 2 g/cm³ and is usually 1.6 g/cm³ and below; The density of transition forms of carbon also does not exceed ~1.5–2 g/cm³.[27] The densities of the foam boxes are lower than the densities of various carbons by 1.2–1.5 times (Table 2.5). This may be due to a closed system of pores whose cavities are inaccessible to liquids in the pycnometer, or imperfect packing of carbon layers and crystallites in the structure of the foam boxes.

TABLE 2.5 Densities of Foams (ρ) Obtained from Compositions Nos. 1–12 at 800°C.

Number composition	ρ, g/cm³	Number composition	ρ, g/cm³
1	1.280	7	1.304
2	1.249	8	1.151
3	1.335	9	1.401
4	1.276	10	1.497
5	1.131	11	1.445
6	1.183	12	1.451

For penokoksov such an indicator as the multiplicity of swelling was studied (Table 2.6), showing a relative increase in the volume of carbon foam in comparison with the initial composition, (Eq. 2.3)

$$K = \frac{V_{fc}}{V_{ek}},$$

(2.3)

where K is the multiplicity of swelling; V_{fc}—effective (external) volume of foam cokes; V_{ek} is the volume of the initial EC.

TABLE 2.6 Multiplicity of Swelling of the Foam Boxes Obtained from Compositions Nos. 1–12 at 800°C

Number composition	K, time	Number composition	K, time
1	23.59	7	31.02
2	30.21	8	38.28
3	18.43	9	33.54
4	38.66	10	9.59
5	33.80	11	20.20
6	30.12	12	12.62

The multiplicity of swelling for the ECCS is approximately —two to three times greater than for the initial epoxy resin with the addition of ammonium polyphosphate (Table 2.6). The maximum value of K is observed for the additives containing nickel and cobalt, approximately three times more than for the composition without additives. The addition of activated carbon also increases the folding rate by almost two times, the addition of graphite powder, on the contrary, slightly decreases—approximately 1.3 times.

To estimate the heat-shielding properties, the coefficient of thermal diffusivity of foams expanded at 800°C was calculated by the expression.[29]

$$a = \frac{\lambda}{C_m \rho_{fc}^*} \tag{2.4}$$

where λ is the thermal conductivity, W/(m · K); C_m is the specific heat, J/(kg · K), ρ_{fc}^* is the apparent density of the foam boxes, taking into account the multiplicity of the swelling

$$\rho_{fc}^* = \frac{m_{fc}}{V_{fc}^*} \tag{2.5}$$

where m_{fc} and V_{fc} are the mass and volume of a swollen foam sample, respectively.

Considering that $m_{fc} = m_0 \cdot \Delta m/100\%$ (where m_0 and Δm (%) are the mass of initial ECs and the loss of mass of ECs after heat treatment at 800°C), $V_{fc} = V_0 K$ (where V_0 and K are the initial volume of ECs and the coefficient of swelling, respectively) and the density of the initial ECs $\rho_0 = m_0/V_0$, we obtain

$$\rho_{fc}^* = \rho_0 \frac{\Delta m}{K100} \tag{2.6}$$

where, $\rho_{fc}, \rho_0,$ g/cm³; Δm, %.

The true and apparent densities of the foam boxes differ by —one to two orders of magnitude (Tables 2.5 and 2.7).

The thermal diffusivity of foams (a) for ECs of nanostructures (Nos. 1–9) is 1.5–2 times higher than for foams taken from ECs Nos. 10–12 (Table 2.8).

The approximate value of thermal diffusivity calculated from polycrystalline graphite average values for polycrystalline graphite at 25°C (density ≈ 1.6 g/cm³, $\lambda \approx 200$ W/(m · K), Cm ≈ 720 J (kg · K)[27]) Is $\sim 1.74 \cdot 10{-}4$ m²/s, that is, approximately two orders of magnitude higher than the values obtained for the foam boxes. The values obtained for foams from ECs of nanostructures are slightly higher (approximately 1.5 times) than for foam boxes Nos. 10–12, which can be explained by the presence in the first small amounts of metals having a high thermal diffusivity. Among penokoksov, obtained from ECs of nanostructures, the least thermal conductivity is Penocox Nos. 1, 6, 9 (Table 2.8).

TABLE 2.7   True Densities of the Epoxy Compositions (ρ_0) and the Apparent Densities (ρ_{fc}) Obtained from them at 800°C of the Foam.

Number composition	Δm, %	ρ_0, г/см³	$\rho_{fc}^{\cdot}$, г/см³	Number composition	Δm, %	ρ_0, г/см³	$\rho_{fc}^{\cdot}$, г/см³
1	88.2	1.254	0.046	7	87.7	1.248	0.035
2	88.7	1.250	0.037	8	87.6	1.245	0.029
3	89.1	1.257	0.061	9	88.1	1.249	0.032
4	87.9	1.242	0.028	10	96.0	1.224	0.122
5	87.9	1.240	0.032	11	96.4	1.211	0.058
6	88.3	1.242	0.036	12	97.3	1.219	0.094

TABLE 2.8   Coefficients of Thermal Diffusivity of Foams, Obtained from Compositions Nos. 1–12 at 800°C.

Number composition	$a\ 10^{-6}$, m²/s	Number composition	$a\ 10^{-6}$, m²/s
1	1.33	7	1.91
2	1.66	8	2.25
3	1.37	9	1.42
4	2.16	10	0.53
5	1.96	11	1.16
6	1.55	12	0.78

The compressive strength of the foams (Table 2.9) was determined by the above-described procedure.

The strength of the foam boxes after the introduction of additives (Nos. 1–9) is significantly increased (in ~2.5–3 times), whereas the introduction of graphite powder does not lead to an increase in the strength of the foam boxes, and the activated carbon to an insignificant (~1.25 times) increase (Table 2.9).

Thus, the introduction of heat treatment products of gels significantly increases the mechanical strength of the foam cages of ECs. Previously, with the example of fullerenes C60, it was shown[30] that carbon nanostructures are better bound by polymers containing aromatic rings (polystyrene) than do not contain them (polymethyl methacrylate). Similarly, ONT in the composition of epoxydian resin leads to a certain decrease in its flexural strength.[31] In accordance with this, an increase in the strength of foams with nanostructures can be attributed to this fact, since fragments of aromatic macromolecules appear to be present in the structure of the foam boxes. The strength of foam boxes is an important indicator, since the flame retardant effect, in contrast to the heat-shielding effect, also depends on the resistance to breaking of the foam boxes under the action of the flame front and the mechanical stresses arising in the intumescent coating.

TABLE 2.9 Strength of Foam Cokes.

Number composition	F, g/cm^2	Number composition	F, g/cm^2
1	12.10	7	12.12
2	12.39	8	11.87
3	12.85	9	13.21
4	14.38	10	4.95
5	14.03	11	3.99
6	13.02	12	4.07

Analyzing the strength data of the foam boxes on compression and the multiplicity of swelling of the compositions, it can be stated that the best in these indices are the products of heat treatment of gels of the composition PVA: $NiCl_2$ 1:1; 5:1 and PVA: $CoCl_2$ 20:1—for epoxy-containing nanostructures containing them, the maximum strength of the foam boxes and the multiplicity of expansion are observed.

2.3.1.2 INVESTIGATION OF THE STRUCTURE OF FOAM COKS BY X-RAY PHASE ANALYSIS

The structure of the foam boxes was investigated by X-ray diffraction in the angular interval 2θ 5–90°. Samples were studied on radiation of equal intensity. Intensive wide peaks of graphite-like carbon are observed at all spectra at $2\theta \approx 23°$ (peak 1) and for $2\theta \approx 31$–32° and at (peak 2), the relative intensities of which vary insignificantly for different samples (Figs. 2.10 and 2.11). The intensities of peak 1 for systems containing an additive with copper, nickel, and cobalt are approximately the same (Fig. 2.10a and c).

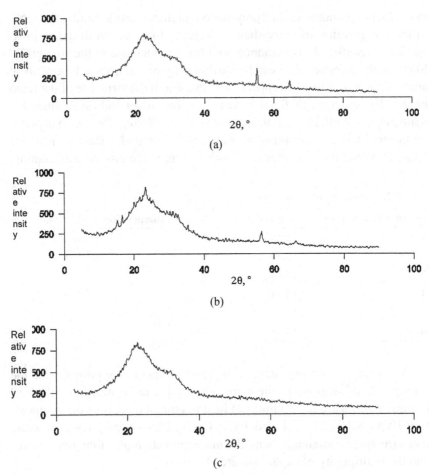

FIGURE 2.10 Diffractograms of foam boxes obtained from compositions Nos. 2 (a), 5 (b), and 8 (c).

At the same time, for systems with additions of graphite and activated carbon, the intensities are somewhat lower, and for system No. 12 (ED with ammonium polyphosphate), the graphite-like peak is approximately 2.5 times less intense than for systems Nos. 2, 5, and 8 (Fig. 2.11a and c). By the position of the peaks of 1 and 2, the corresponding interplanar distances were calculated, and the intensity of the peaks with respect to the sample containing activated carbon was also measured (Table 2.10).

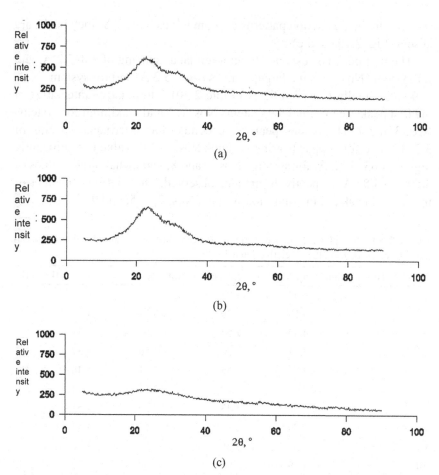

FIGURE 2.11 Diffractograms of foam cokes, obtained from compositions Nos. 10 (a), 11 (b), and 12 (c).

In the diffraction patterns of the foam boxes Nos. 2, 5 (Fig. 2.10a and b), two peaks corresponding to metallic Cu and β-Ni (angles 2θ ~55.5, 64.6 and 56.9, 66.7°, respectively). In the X-ray boxes No. 8 by the XRD method, metallic cobalt was not detected, the corresponding peaks are absent. Apparently, this is due to a relatively smaller amount of cobalt-containing additives. The diffractogram of the foam boxes No. 5 contains eight small clear peaks in the range of angles 18–26° (Fig. 2.10b) belonging to the nickel (II) chloride hydrate with different contents of the crystallization water, probably formed during sample preparation for the survey. On the diffraction patterns of foam boxes Nos. 2, 8, such peaks are absent (Fig. 2.10a and c).

The first peak corresponds to an interplanar spacing of 4.8–5.2 Å, and for systems Nos. 1–3, the largest values ~4.9–5.2 Å, and for systems Nos. 4–9—the smallest: 4.77–4.92 Å (Table 2.10). The relative intensities of the first peak are minimal for systems Nos. 1, 6 and maximum for systems Nos. 3 and 7. The second peak corresponds to an interplanar spacing of 3.5–3.7, and for copper-bearing systems Nos. 1–3 its value is significantly higher: 3.63–3.72 Å; for systems Nos. 7 and 8, somewhat larger values of 3.59 and 3.55 Å, respectively, are also observed. The relative intensities of the second peak are maximal for systems Nos. 2, 3, 5, and 9.

TABLE 2.10 Interplanar Distances Calculated from the Peaks 1 and 2 of the Graphite-like Phase ($d1$, $d2$) and their Relative Intensities (I_1, I_2).

Number composition	d_1, Å	I_1, relative unit	d_2, Å	I_2, relative unit
1	5.09	1.03	3.72	1.39
2	5.19	1.29	3.72	1.57
3	4.89	1.54	3.63	1.83
4	4.92	1.26	3.50	1.37
5	4.75	1.46	3.51	1.48
6	4.77	1.09	3.51	1.44
7	4.89	1.43	3.59	1.51
8	4.79	1.39	3.55	1.52
9	4.88	1.36	3.51	1.75
10	4.79	0.95	3.50	1.13
11	4.77	1	3.50	1
12	4.47	0.29	–	–

Figure 2.12 allows to establish the correlation of strength of foam boxes with its structural characteristics.

The greatest strength (Fig. 2.12a) has foam boxes Nos. 1, 4, 5, 9 and the smallest—Nos. 10–12. As can be seen from Figure 2.12b, for peak 1, the parameters do not correlate in any way with the strengths of the foam boxes. At the same time, a simple connection can be established between the parameters of peak 2 and strengths (Fig. 2.12c). The strength of the foam boxes in the series of samples Nos. 1–3 decreases (Fig. 2.12a), the interplanar spacing is the same for Nos. 1, 2 (3.72 Å), and slightly less for Nos. 3 (3.63 Å); the intensity of the peaks increases monotonically. The strength of foam boxes Nos. 4–6 also decreases, the maximum value is observed for Nos. 4. The value of the interplanar distance for Nos. 4 is somewhat smaller (3.50 Å) than for Nos. 5, 6 (3.51 Å). The lowest intensity of the peak is observed for Nos. 4, the largest for No. 5, and for No. 6, the average value is close to No. 5. For the foam boxes Nos. 7, 8, the strengths are approximately the same, but for No. 9 it is much higher. At the same time, the values of d and intensities for Nos.7, 8 are also close (d 3.59 and 3.55 Å, respectively). For Nos. 9, the value of d is minimal (3.51 Å), and the intensity is maximum in series Nos. 7–9.

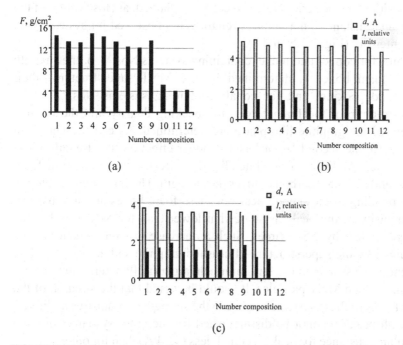

FIGURE 2.12 Strength of the foam boxes obtained from compositions Nos. 1–12 at 800°C (a), interplanar distances and relative intensities for the peaks 1 (b) and 2 (c).

Thus, in the general case, for systems Nos. 7–9 containing cobalt, the strength of the foam cages is higher, that is, the smaller the interplanar distance corresponding to the graphite-like peak 2, the higher is its intensity. For systems Nos. 1–3 containing copper, there is an opposite trend. The observed dependence of the strength of foam boxes for systems Nos.

7–9 seems to be natural since it increases with increasing packing density of the layers in the foam boxes (decrease in the interplanar distance) and the number of sections with such a structure (growth of the relative intensity of the peak). For systems Nos. 1–3, the decrease in the strength of the foam can be explained by the change in the superstructure of the garnet carbonization product—with an increase in the relative amount of PVA in the gels, the MNT structure in the product changes from the interlaced mesh (Fig. 2.13d) to direct single MWTs (Fig. 2.14c). Accordingly, the mechanical strength of the foam boxes containing MNT in the form of a woven "network" exceeds the strength of the foam boxes containing direct single MNTs. Similarly, the strength of foam-coxes of nickel-containing compositions in the series Nos. 4–6 can be explained: at close values of the interplanar distances, maximum strength is not observed for systems with peak intensity (Fig. 2.12a and c).

Thus, for copper and nickel containing systems Nos. 1–6, the strength of the foam boxes is determined by the MNT superstructure—their shape and packing; and for cobalt containing systems Nos. 7–9, the microstructure of the foam boxes: the packing density and the relative content of the graphite-like phase are determined. As can be seen from Table 2.10, the second factor has a stronger effect on the strength of the foam boxes. Addition of graphite slightly increases the strength of foam boxes, and activated carbon reduces the strength. The interplanar distances corresponding to the second peak for these foam boxes are minimal (6), the intensity of the peaks is somewhat lower than for Nos. 1–9, but the strength is less by 2.5–3 times. Such a decrease in strength can also be explained by the superstructure of crushed graphite and activated carbon present in the system in the form of much larger (millimeters and micrometer sizes) than MNT particles. It should be noted that the strength of the foam boxes of these systems is close to the strength of foam-free additives, although peak 2 cannot be distinguished for the latter system, while the interplanar distance for peak 1 is much less (4.74 Å) than for other systems (4.8–5.2 Å).

As it was shown, the minimum values of thermal diffusivity are detected by foam boxes Nos. 1, 6, 9, 10–12 (Table 2.8). At the same time, the maximum strength of the foam boxes in combination with the greatest multiplicity of expansion has foam boxes from ECs of nanostructures, especially Nos. 4, 5, 9, exceeding in these parameters of system Nos. 10–12 by 1.5–3 times. Thus, the foam cages obtained from ECs with additives of nanoscale carbonization products have greater strength than ECs without additives (or containing activated carbon and graphite) in combination with low thermal diffusivity coefficients comparable with those obtained for foam cages from epoxy nanostructure compositions Nos. 10–12. Therefore, it can be argued that the foam boxes obtained from the ECs Nos. 1–9 have, in general, the best fireproofing properties.

2.4 INFLUENCE OF CARBON NANOTUBES ON THE PROPERTY OF FOAM COKES

2.4.1 STUDY OF THE EFFECT OF CARBON-CONTAINING METAL NANOSTRUCTURES BASED ON PHENANTHRENE ON THE PROPERTIES OF FOAMS

The physical and chemical structure of the surface of materials determines many practically important macroscopic properties of materials and products. It is important to take into account that the mechanical transfer of the characteristics of the volume of materials under study to the properties of their surface may lead to a misunderstanding of the phenomena occurring at the interface. Even with respect to such relatively simple systems as inorganic crystals, judging the properties of their surface layers by extrapolating the volume characteristics is not always correct. It is known that the adsorption properties of various crystalline faces can vary greatly.[16] Especially, the peculiarities of the structure of surface layers should be taken into account when studying such complex objects as polymeric materials and biological systems.

As a polymeric base, an epoxy–epoxy resin of ED-20 grade was used. The epoxy resin was mixed with the ammonium polyphosphate pre-ground in a porcelain mortar and cured by PEPA. Separate samples of uncured modified polyphosphate epoxy resin were mixed with manganese dioxide and calcium borate to form mineral network structures in the surface layers, and also mixed with five parts by weight with products of carbonization of PVA gels with manganese chlorides, phenanthrene with nickel chlorides (FA (Ni)), and nickel (the nickel content is two times larger—FA2 (Ni)) (II) (nanostructures), and also cured with polyethylene polyamine at room temperature for 7 days. Before adding to the ECs, the solid components were ground in a porcelain mortar to a dust state (Table 2.11).

TABLE 2.11   Composition of Intumescent.

Components	Composition, parts by weight											
	1	1b	2	3	3b	4	4b	5	5b	6	7	8
ED-20	10.0	10.0	10.0	10.0	10.0	10.0	10.0	10.0	10.0	10.0	10.0	10.0
PFA	2.0	3.0	2.0	2.0	3.0	2.0	3.0	2.0	3.0	3.0	3.0	3.0
T (Ni)												0.8
T (Cr)	0.4	0.4	0.4							0.4	0.8	
Calcium borate	0.4	0.4		0.4	0.4	0.4	0.4					
MnO$_2$				0.4	0.4					0.4	0.8	0.8
PP	1.5	1.5	1.5	1.5	1.5	1.5	1.5	1.5	5	1.5	1.5	1.5

The cured compositions were thermally treated in the air in a flame of a gas burner until the formation of foam boxes (T ~ 800°C) was completed. To investigate the structure of the obtained foam-coxes, atomic force microscopy (AFM), local power spectroscopy (A-mode), and XPS were used.

According to the laws of thermodynamics, any system tends to form a surface with a minimum free surface energy. In the formation of the surface of solids, this leads to the fact that groups or components that provide a minimum interfacial energy are localized at the phase boundary. The process of reducing the interfacial energy can occur in two ways.

First, if the components of the material have sufficient mobility, the components migrating to the surface of the material ensure a minimum gradient of the surface energy at the interface. In particular, functional groups, macromolecules or components with a minimum surface energy are concentrated at the polymer–air interface; while concentrating, the most polar components or functional groups are desirable at the interface of the polymer with air in terms of the favorable energy state.

Second, if the material is already formed, then the decrease in the surface energy gradient at the phase interface occurs by sorption of gaseous and liquid components from the environment on the surface of the material.[32]

The last decade in experimental physics is characterized by the intensive development of fundamentally new microscopic methods for studying the surface with nanometer and atomic spatial resolution. At present, they are united under a common name—scanning probe microscopy.[33]

The appearance in the last 15 years of new probe methods for studying the surface, such as scanning tunneling microscopy (STM) and AFM, made it possible to study topography and various surface characteristics in the limit with atomic resolution. The most informative in this respect for the surface physicochemistry (especially for nonconducting objects) was the AFM, which, in addition to topography, obtained a number of power local characteristics of the surface (rigidity, adhesion, friction, etc.). As a rule, not only the topography of the object contributes to the images obtained by the STM or AFM methods, but also the effects of the force interaction of the needle with the surface, which can give information about the local distribution of functional groups at the molecular level, directly examining interactions between molecules or molecular structures with a high lateral resolution, which can be a new unique source of information about local (including chemical) surface properties.

Both internal and external layers of foam boxes were studied. The study was carried out on a P4-Solver device manufactured by NT-MTD. The flint probes of the same firm of grade CS12 were used. The strength of the coatings was assessed by the minimum force of pressing the needle against the specimen, at which the bands remain on the coating after multiple scanning of the selected area. The clamping force was adjusted by changing the set point meter. The probe was scanning at a speed of 6 μm/s, and a scan site of 1 μm^2 was scanned. A step load (Fu) was applied: (1) Fu $= 1.35$ μN, (2) Fu $= 2.7$ μN, (3) Fu $= 4.05$ μN. The change in the polarity of the coating, depending on the formulation, the outer and inner surfaces of the foam coke, was evaluated from the change in the interaction force between the sample and the probe by removing the "force-distance" curves.[34,35] In fact, on these curves along the Y axis, the amount of deviation of the console (in nanometers) from its position in the unloaded state when the needle is approaching and withdrawing from the sample is displayed. Using calibration on the test grid, you can translate the values deposited along the Y axis into units of length. For this device, 1 nA corresponds to 150 nm. When the needle is removed from the sample (after contacting them) due to the action of adhesion forces 35, the needle follows the sample until the elastic force of the console exceeds the force of interaction between the needle and the sample. At this point, the needle is torn from the sample, and the console returns to the original undeformed state.

A change in the chemical structure of the surface and volume of intumescent systems during pyrolysis was carried out using XPS on an ES-2401 spectrometer (MgKα radiation) and a magnetic photoelectron spectrometer (AlKN radiation). In turn, the samples were pyrolyzed according to the following procedure: the sample was placed in a molybdenum cuvette and then introduced into a tubular quartz furnace. The pyrolysis temperature was set by an electronic control unit with feedback. The accuracy of maintaining the temperature is 2%. Pyrolysis was carried out stepwise in an atmosphere of nitrogen, free from oxygen impurities. The pyrolysis temperature was selected from differential thermal analysis and TG data.

Since the formation of foam blocks can be represented as successive stages during heating of the compositions, it can be assumed that due to the difference in thermal expansion near the ammonium polyphosphate particles, microcracks are formed, into which ammonia and water vapor enter and simultaneously form polyphosphoric acid at the walls. The latter can act as a dehydrating agent and as a carbonation stimulator. Then we can expect that a carbon (close to a graphite-like) layer appears on the inner surface of the bubble formed. According to the XPS spectra, it is indeed at a certain stage of temperature heating that a graphite-like layer is formed on the inner surface of the bubble (Ecb = 283.5–284 eV). Since the surface and boundary layers of the cured resin, the products of its thermolysis, as well as the pyrolysis residues of the materials were studied by X-ray spectroscopy, the results of studying the initial resin should be considered.

When the sample is heated to the pyrolysis start temperature (533 K), phosphorus-containing substances are released to the surface of the sample (ammonium salts of polyphosphoric acid are formed). It should be noted that the temperature of the beginning of the decomposition of PFA is 463 K. Judging by the fact that the nitrogen line with a binding energy N1s 402.1 eV is preserved in the spectra up to a temperature of 573 K, the decomposition reaction has a more complex character. It is possible that complete deamination of PFA, and therefore its complete destruction into the polymer, is difficult, due to the high density of the polymer and its degradation products up to high temperatures. The obtained data are confirmed by determining the temperature dependence of the heat capacity of the studied materials.

The output of phosphorus-containing products is accompanied by an increase in the phosphorus concentration on the surface, as well as a redistribution of the intensity of the nitrogen lines in the X-ray photoelectron spectrum (an increase in the higher-energy line). When the temperature in the boundary layers increases, the relative amount of C-OR-groups in the volume of the material decreases along with the carboxyl groups.

Intensive formation of liquid pyrolysis products begins at a temperature of 573 K (the pyrolysis start temperature), as evidenced by a sharp increase in the intensity of the C1s line attributed to the CH groups.

Carbon products accumulate on the surface of the samples, masking all the other fragments of pyrolysis products, therefore it can be assumed that the process of coke formation begins at a temperature of 573 K. No more than 7.24% of carbon is found on the cleavage surface of the samples. Apparently, the interaction of PFA with the polymer continues at this temperature, since the intensity of the nitrogen line with E = 402.1 eV decreases along with a decrease in the binding energy of phosphorus.

In comparison with the results obtained, the introduction of nickel-containing tubulenes into the composition at the ratio of PFA to Ni-T 10:1 leads to an increase in carbon–carbon and carbon–metal groupings by almost three times (Fig. 2.13).

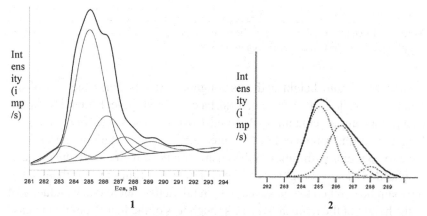

FIGURE 2.13   1: Results of X-ray spectroscopy on the 1Cs line in the study of an EC containing: a—ammonium polyphosphate and T-Ni; b—ammonium polyphosphate (beginning); 2: results of X-ray spectroscopy on the 1Cs line in the study of an EC containing: a—ammonium polyphosphate and T-Ni; b—ammonium polyphosphate (ending).

Analysis of AFM images in normal forces has shown that the topography of the outer and inner layers of foam boxes of different intumescent systems based on the epoxy-resin ED-20 varies significantly. Even significant differences in the topography of the outer and inner layers of the same foam box. This is evidenced by the value of the maximum scatter of heights (Sa), root-mean-square (rms) roughness (Sg) (Table 2.12).

TABLE 2.12   Results of the Study of Cokes by the Atomic Force Microscopy Method.

Sample	Properties			
	Strength of adhesion, Fa, nN	Destructive load, Fd, nN	Statistics	
			Sweep of heights, Sa, nm	Root-mean-square roughness, Sg, nm
No. 1 surface	680	>3.2	7	8
No. 1 volume	20	>3.2	9	10
No. 2 surface	8	>3.2	18	21
No. 2 volume	35	>3.2	59	69
No. 3 surface	340	0.5	9	11
No. 3 volume	450	>3.2	76	95
No. 4 surface	90	1.8	20	24
No. 4 volume	935	3.2	10	12
No. 5 surface	755	0.5	16	19
No. 5 volume	214	3.2	41	49
No. 6 surface	135	0.5	34	39
No. 6 volume	90	>3.2	98	57

Note: Fa: force of adhesion of the probe to the sample, nN; Fd: normal load at which the sample surface is destroyed during scanning, nN.

The maximum height and rms roughness for the outer surface of the foam box are less than the inner surface. This is typical for all samples, except for samples containing PFA and T (Mn). Close values of the range of heights and rms roughness of the outer surface and in the volume of foam coke for samples with MnO_2. When using tubulene with nickel, the range of heights and the rms roughness of the volume is 3.3 times greater than that of the foam box. When the nickel content in tubulene is increased by a factor of 3, the height of the foam and the rms roughness of the foam volume increase by 8.4 times relative to the surface. When using T (Mn) as the modifying additive, the height range and the rms roughness for the outer surface of the foam box is 2.0 times the volume. For sample No. 6, T (Mn) was used as the modifying additive, which was prepared on the basis of PVA. For this sample, the height sweep and the rms roughness for the outer surface of the foam box are 2.9 times smaller than in volume (Fig. 2.13).

When examining the strength of the outer and inner surfaces, it is determined that the outer surface of sample No. 1 did not collapse under maximum loads. The internal surface of sample No. 1 was destroyed at a load of Fu=2.7 µN (Table 2.12). The internal surface of sample No. 2 did not collapse under maximum load. Based on the test results, it can be concluded that the formation of the foam box begins with the inner surface of the foam coke. This effect can be explained by the fact that on the inner surface of the sample, according to the obtained data of X-ray spectroscopy, there is an increased content of polyphosphoric acid, whereby stimulation of the processes of foam coking occurs.

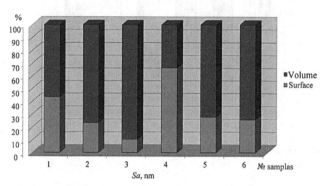

FIGURE 2.14 **(See color insert.)** Height of the outer and inner surfaces of the foam box.

The adhesion strength is manifested on the curve F (z) as a step ΔD, the value of which is proportional to the force of interaction of the needle with the sample. By the value of ΔD, using the values of the force constant for this console, it is possible to calculate the value of the force F needed to detach the needle from the sample:

$$F = K\Delta D$$

where K is the power constant of the console; ΔD is its deflection.

Spectroscopy was carried out at nine points of the AFM image of the coating surface, and then the average value of ΔD and the adhesion strength were calculated. It has been established that the force of the needle detachment from the outer surface of the foam coke of sample No. 1 is 2.97 µN, from the inner surface of sample No. 1 the tearing force is equal to 4.12 µN. From the outer surface of the foam box of sample No. 2, the needle does not break (there is not enough rigidity of the beam), the adhesion strength is very high. The force of separation of the needle from the inner surface of sample No. 2 is 4.12 µN (Table 2.12, Figure 2.15).

Since a needle with a polar surface (silicon oxides) is used, an increase in the adhesion strength to the surface of the foam box indicates an increase in the polarity of the surface,[32] hence the polarity of the foam box surface changes with increasing PFA concentration and the migration of P, N containing to the foam box surface increases.

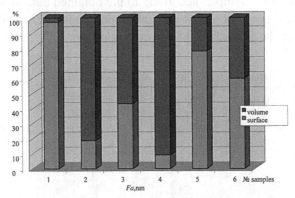

FIGURE 2.15 **(See color insert.)** Strength of the probe's adhesion to the outer and inner surfaces of the foam box.

In the photograph of sample No. 1, the pores are located along the surface, therefore the range of the surface and volume heights is close (Figs. 2.16 and 2.17).

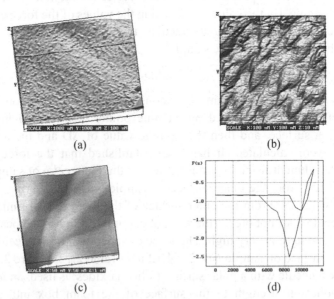

FIGURE 2.16 Sample No. 1: a, b, c—an image of the surface of coke. At different magnifications; d is the characteristic form of the force-distance curve.

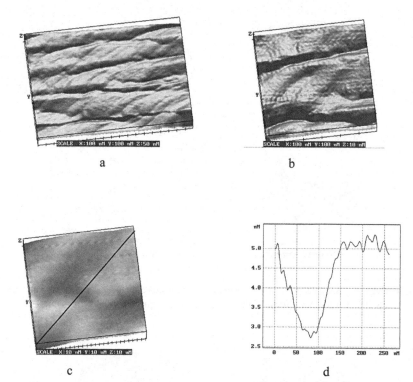

a   b

c   d

FIGURE 2.17 Sample No. 1: a, b, c—an image of the surface of coke. At different magnifications; d is a characteristic form of the curve "force-distance."

The photograph of sample No. 2 shows the ordered structure of the foam coke in the volume (Figs. 2.18 and 2.19). The shape of the blisters is different in all samples. The structure of the foam box is layered. It is possible that with the help of these edges bubbles form. Bubbles have an elongated shape upwards (samples Nos. 2–4) (Fig. 2.18–2.23) and in along the surface of samples Nos. 1, 5, 6 (Fig. 2.16, 2.17, 2.24–2.26).

The introduction of such structures allows one to change or influence not only the structure of the foam box but also the structure of the foam box affects the shape and arrangement of the bubbles.

In sample No. 1, the breaking load is greater than 3.2 nH surface and volume. The structure of the surface and volume are close. When T (Ni) is introduced into sample No. 2, the same regularities are present, but the structure is more ordered than sample No. 1. The surface and volume structure are close.

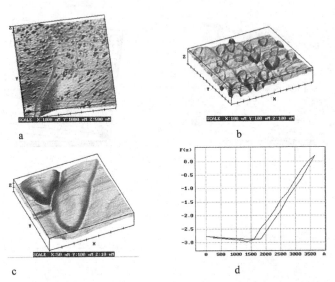

FIGURE 2.18   Sample No. 2: a, b, c—an image of the coke surface for a different increase rates; d is the characteristic form of the "force-distance" curve for the same sample.

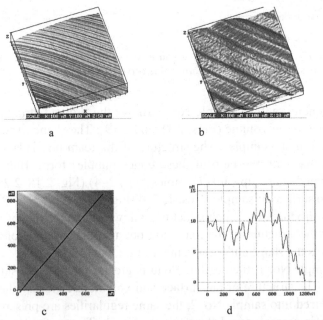

FIGURE 2.19   Sample No. 2: a, b, c—an image of three-dimensional layers of coke at different magnifications; d is the profile of the section of the surface (on b is shown by a line).

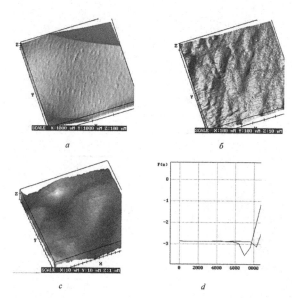

FIGURE 2.20 Sample No. 3: a, b, c—an image of the coke surface for a different Increase; d is the characteristic form of the "force-distance" curve for the same sample.

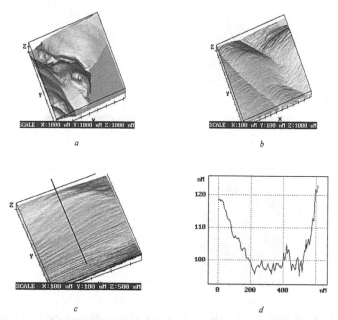

FIGURE 2.21 Sample No. 3: a, b, c—an image of three-dimensional layers of coke at different magnifications; d is the profile of the section of the surface (on b is shown by a line).

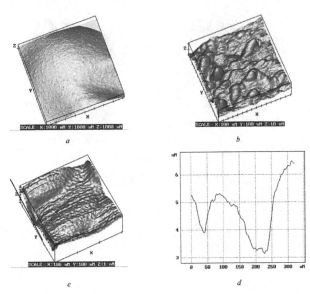

FIGURE 2.22 Sample No. 4: a, b, c—an image of the coke surface for a different Increase; d is the characteristic form of the "force-distance" curve for the same sample.

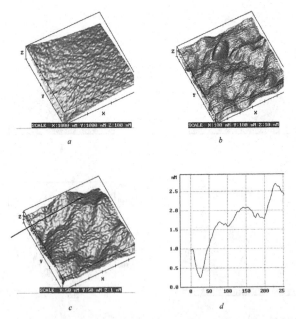

FIGURE 2.23 Sample No. 4: a, b, c—an image of three-dimensional layers of coke at different magnifications; d is the profile of the section of the surface (on b is shown by a line).

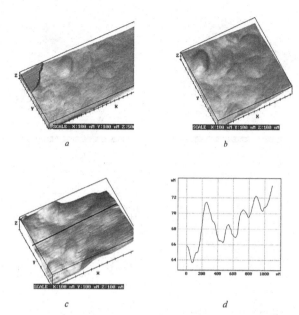

FIGURE 2.24 Sample No. 5: a, b, c—an image of three-dimensional layers of coke at different magnifications; d is the profile of the section of the surface (on b is shown by a line).

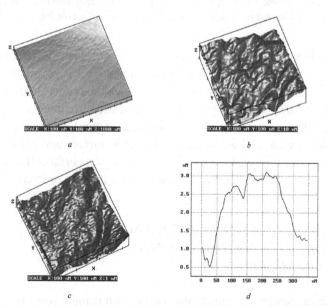

FIGURE 2.25 Sample No. 6: a, b, c—an image of the coke surface for a different increase rates; d is the characteristic form of the "force-distance" curve for the same sample.

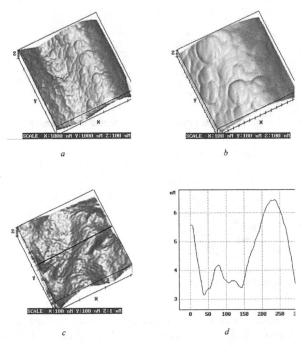

FIGURE 2.26   Sample No. 6: a, b, c—an image of three-dimensional layers of coke at different magnifications; d is the profile of the section of the surface (on b is shown by a line).

In sample No. 3, the difference in the orientation of the structure of the bulk layer of the foam box compared with the surface layer is associated with an increase in the nickel content in the tubule or nanotube (Figs. 2.20 and 2.21).

The use of another metal (manganese) as a structurant makes it possible to obtain a less oriented (ordered) structure of the foam box relative to sample No. 2, where Ni was used, but closer in surface and volume orientation, therefore the breaking loads are close (for a surface of 1.8 nN and for Volume 3.2 nH), but in the volume of foam is more ordered.

2.4.2 INVESTIGATION OF THE PROPERTIES OF INTUMESCENT COATINGS USING A SCANNING ELECTRON MICROSCOPE

Surface investigation of objects was carried out using a scanning electron microscope REM-100U with an accelerating voltage of 30 kV for various magnifications by scanning with an electronic probe over the surface of the object under study.

A sample on a metal substrate was fixed with a conductive adhesive. The obtained images of the investigated surface were fixed onto a photographic film.

Each sample was examined from two sides—external and internal. Investigation of the distribution of nickel, calcium, and phosphorus over the surface of objects was carried out with the help of a micro-ion attachment to the REM-100U. LiF, CAR, and PET were used as crystal analyzers.

The foam boxes were obtained by burning in a flame of a gas burner at a temperature of 800°C with a modified ED-20. The residence time of the sample in the flame of the burner was determined by the end of the process of foam coking. Compositions of the intumescent are given in Table 2.13.

TABLE 2.13 Compositions of Intumescent.

Components	Composition, mass (%)		
	1	2	3
ED-20	10.0	10.0	10.0
PFA	3.0	3.0	3.0
T(Ni)	–	0.4	–
Calcium borate	–	–	0.8
PP	1.5	1.5	1.5

The surface structure of the samples is shown in Figures 2.27–2.34. Studies of the distribution of nickel, calcium, and phosphorus have not yielded positive results. A negative result can be related to the nature of the object of study and the capabilities of the device.

Photographs of foams were obtained at various magnifications. The surface of the foam box formed during the combustion of a composition containing, in addition to PFA, carbon–nickel-containing nanostructures is a set of bubbles whose dimensions are approximately the same and relatively small (Fig. 2.27). Inside, the foam box has a porous structure with very thin walls and high air content in the inter-porous space (Figs. 2.28 and 2.29).

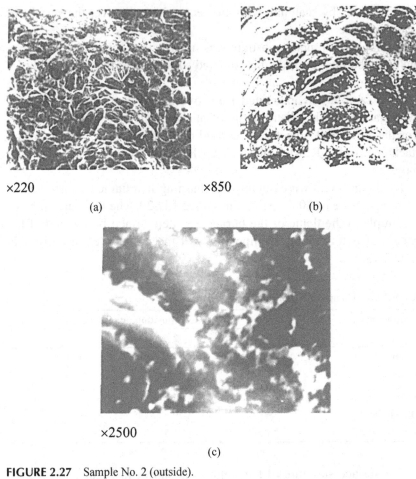

×220

(a)

×850

(b)

×2500

(c)

FIGURE 2.27 Sample No. 2 (outside).

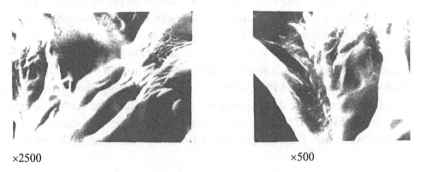

×2500

×500

FIGURE 2.28 Sample No. 2 (inside).

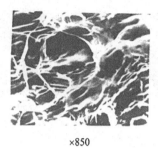

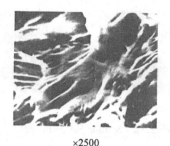

×850   ×2500

FIGURE 2.29 Sample No. 2 (inside).

The surface of the foam coke of composition 3 is similar to the surface of foam coke of composition 2 (Fig. 2.30), although the bubble size is much higher. The internal structure also has significant differences (Fig. 2.31). The porous structure of the inner part is preserved, but the walls of the pores are much thicker. The surface of the foam coke of composition 1 (Fig. 2.33) is even more distinguished by an increase in the size of the bubbles relative to the foam cages of compositions 2, 3 (Fig. 2.31). The porous structure of the inner part is preserved (Fig. 2.34), but the walls of the pores are much thicker, relative to the inner part of the foam coke of the composition 3.

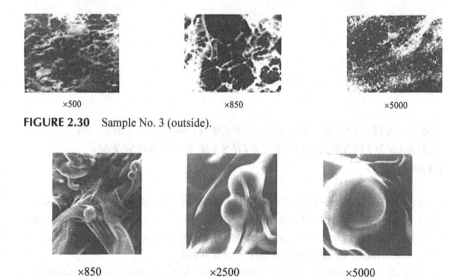

×500   ×850   ×5000

FIGURE 2.30 Sample No. 3 (outside).

×850   ×2500   ×5000

FIGURE 2.31 Sample No. 3 (inside).

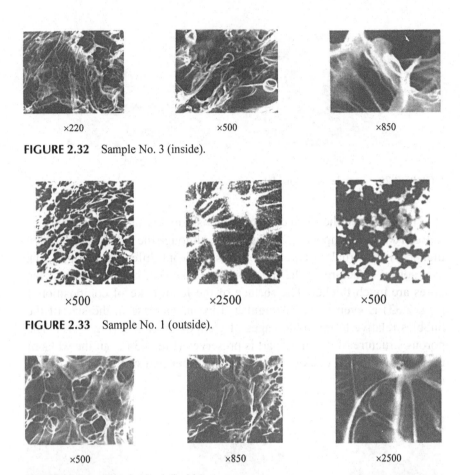

×220   ×500   ×850

FIGURE 2.32   Sample No. 3 (inside).

×500   ×2500   ×500

FIGURE 2.33   Sample No. 1 (outside).

×500   ×850   ×2500

FIGURE 2.34   Sample No. 1 (inside).

2.4.3 STUDY OF THE EFFECT OF TEMPERATURE ON THERMOPHYSICAL PROPERTIES OF INTUMESCENT COATINGS

Energy transfer processes play a fundamental role in solid-state physics. That is why when describing the properties of a solid body, it is necessary to consider the effect of temperature on thermal conductivity, as well as other thermophysical characteristics of a solid.

The phenomenon of thermal conductivity in many respects is analogous to the other most important kinetic phenomenon—electrical conductivity.

However, it should be noted that the picture of energy transfer in solids is much more complicated than the phenomenon of charge transfer. The fact is that the electrical conductivity is realized only by the electron system, as for energy transfer, it can also be realized by the phonon system, therefore the phenomenon of thermal conductivity is characterized by twice as large as the electrical conductivity by the number of relaxation mechanisms (phonon scattering by electrons, phonon–phonon interaction, and phonon scattering Impurities). Each of the mechanisms is characterized by its temperature dependence, and the overall picture is very complicated. In various situations, determined by the structure of matter, temperature, and so forth, a basic role is usually played by some of the relaxation mechanisms.[36]

There are several mechanisms that cause thermal conductivity in solids and many processes that limit the effectiveness of each mechanism. In non-metals, heat is transferred by thermal vibrations of atoms. In simple metals, such oscillations give some contribution to the thermal conductivity, but the value of the thermal conductivity is almost completely determined by the electrons. It does not follow from this that the thermal conductivities of these two types of solids should be very different in magnitude, as is the case for their electrical conductivity. However, the dependences of the thermal conductivity on temperature and the concentration of defects in the samples differ substantially in both cases.

The studies were carried out with the compositions of the following formulations (Table 2.14).

TABLE 2.14 Formulation of Compositions for Intumescent Coating.

Components	Composition, parts by weight											
	1	1b	2	3	3b	4	4b	5	5b	6	7	8
ED-20	10.0	10.0	10.0	10.0	10.0	10.0	10.0	10.0	10.0	10.0	10.0	10.0
PFA	2.0	3.0	2.0	2.0	3.0	2.0	3.0	2.0	3.0	3.0	3.0	3.0
T (Ni)												0.8
T (Cr)	0.4	0.4	0.4							0.4	0.8	
Calcium borate	0.4	0.4		0.4	0.4	0.4	0.4					
MnO$_2$				0.4	0.4					0.4	0.8	0.8
PP	1.5	1.5	1.5	1.5	1.5	1.5	1.5	1.5	5	1.5	1.5	1.5

In comparison with the results obtained, the introduction of nickel-containing tubulenes into the composition at a ratio of PFA to Ni-T of

10:1 leads to an increase in the carbon–carbon and carbon–metal groups by almost three times (Fig. 2.13). An increase in the content of carbon products in the foam box leads to a significant change in the character of the heat capacity versus temperature (Fig. 2.35). First, the heat capacity of the compositions modified by structurants is three to seven times higher than the heat capacity of the composition containing only ammonium polyphosphate, which is explained by the different degree of structure formation of the compositions. Second, the heat capacity of the composition containing PFA sharply increases by a factor of 10–11 at temperatures in the region of 373–403 and 430–480 K, which is apparently due to the release of vapors of water and ammonia into the bubbles formed and the pressure in the bubbles.

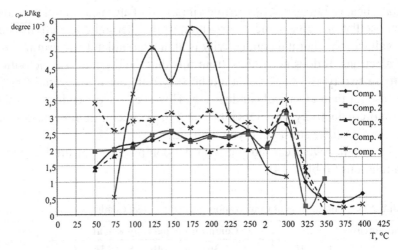

FIGURE 2.35 **(See color insert.)** Dependence of heat capacity on temperature. Comp. nos. 1–5.

Unlike the picture presented in the modified compositions, the heat capacity changes without significant jumps, which is caused by a more gentle flow of the gassing process, and when calcium borate is introduced into the calcium–phosphate–borate composition, the variations in heat capacity are practically insignificant. In this case, it is possible to note more significant changes in the heat capacity (by a factor of 2–3) with the introduction of manganese dioxide along with calcium borate. This is explained, in our opinion, by competing processes of thermal destruction of samples containing manganese dioxide with the formation of volatile inflammable products, the degree of swelling in this composition is 16 times, and the flash point is 463 K.

Data on thermophysical characteristics and the study of X-ray spectra are in good agreement with the results obtained in the study of flash temperatures, the degree of swelling, and the strength of cokes. Below are the final and initial values of the heat capacity of the compositions in the temperature range 323–573 K, their degree of expansion and the flash point.

From the data given, it follows that in the compositions in which manganese dioxide is introduced, in most cases the flashpoint decreases, which can mean the formation of sufficient concentrations of combustible gases at low temperatures. At the same time, the resulting coke, when removed from the flame, lost its shape with a decrease in the degree of swelling (samples 7 and 8) (Table 2.15.)

TABLE 2.15 Thermophysical Parameters of Compositions.

Indicators	Number compositions							
	1	2	3	4	5	6	7	8
$C_p^{\text{к}}/C_p^{\text{н}}$	1.90	1.60	2.30	1.03	11.5			
B, time	4		16	6	12	9	10	7
T_{fl}, K	643		463	583	603	523	593	513

The introduction of chromium- and nickel-containing nanostructures reduces the growth of the heat capacity, reduces the degree of swelling, but leads to an increase in the flashpoint. At the same time, an increase in the ignition time and a decrease in the self-combustion time of samples containing metal-containing nanostructures (tubulenes) Tvsp, K, and/or calcium borate should be noted.

The obtained temperature dependences of the thermal conductivity, shown in Figure 2.36, show that in the temperature range of 100–125°C, the formed nanophases with the help of carbon–metal-containing nano-structures in the polymer matrix help to form more cross-linked structures.

As the density of the polymer increases the thermal conductivity also increases. As the temperature increases, nanophases lead to thermal resistance and the thermal conductivity decreases. In the temperature range up to 300°C, the foam coking process is completed, and a fairly uniform material is obtained. In compositions containing more graphite-like substances (carbon–metal-containing nanostructures), which are both nuclei of coke formation, and regulators of the structure of the formed foam coke.

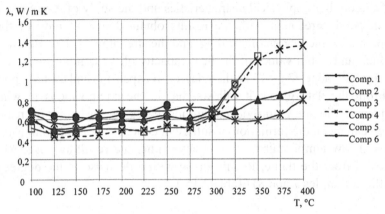

FIGURE 2.36 **(See color insert.)** Dependence of thermal conductivity on temperature. Comp. nos. 1–6.

The influence of small additives (up to 5 wt.%) of carbon–metal-containing nanostructures obtained by the above method carbonization of gels containing PVA and copper (II), nickel (II), and cobalt (II) chlorides in various ratios on the thermophysical properties of the formed carbon foams (foam boxes) at elevated temperatures. The starting epoxydiene resin was modified with ammonium polyphosphate to give it flame-retardant properties. The carbonation products of PVA gels, treated at 450°C for 2 h, were added to the resulting mixtures. Foam cells obtained from a modified resin without an additive and adding carbonization products of PVA gels were studied.

The use of intumescent coatings for thermal protection is caused not only by the low value of their thermal conductivity but also by the large absorption of heat. As a result of decomposition of intumescent compositions, one part is converted to gaseous products, the other remains in the matrix in the form of condensed carbon. When the resin decomposes, its structure changes, leading to the formation of carbon networks and cross-links between them. Thus, as a result of the foam forming of intumescent compositions in the bulk of the material, a secondary carbon skeleton appears whose thermal conductivity is much higher than the thermal conductivity of the filler. Heat transfer in the decomposing material is carried out, in addition to the thermal conductivity of the coke layer, the heat conductivity of the gases in the pores, by radiation in the pores and through the material, and by convection in the pores with the motion of the gases formed.[37,38]

According to the curves presented (Fig. 2.37), it is clear that the compositions with carbon-containing, metal-containing nanostructures containing copper in the ratio 1:20 and 1:5, cobalt 1:5 and 1:1 have the greatest thermal conductivity at 100°C. These values can be explained by more active interaction of nanostructures with a polymer matrix. Therefore, more cross-linked structures are obtained which have a higher density, and hence a higher thermal conductivity. The greatest decrease in thermal conductivity in the temperature range 100–150°C and large changes in the thermal conductivity from 0.09 at 160°C to 1.8 W/m at 210°C has a composition containing carbon–copper-containing nanostructures in the ratio 1:5. The greatest value of thermal conductivity at 400°C has compositions that are modified by carbon nanostructures containing Cu (1:20, 1:1), Ni (1:5), and Co (1:5). Such differences in the values of thermal conductivity can be explained by physicochemical processes occurring at elevated temperatures and the interaction of carbon–metal-containing nanostructures with a polymer matrix.

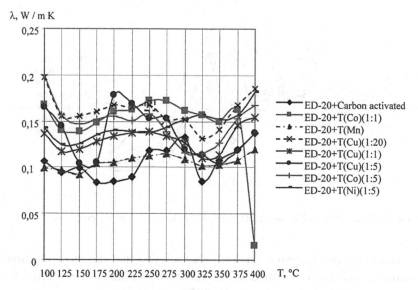

FIGURE 2.37 (See color insert.) Dependence of thermal conductivity on temperature.

Two peaks are observed from the curves of the heat capacity versus temperature (Fig. 2.38), which characterize the main physicochemical processes occurring at temperatures of 200 and 275°C with a slight deviation for individual compositions. The heat capacity decreases by a factor of 1.5–1.8 at temperatures in the region of 100–125°C. At the same time, it is possible to note more significant increases in heat capacity by 0.5–1.8 times at temperatures in the range of 125–200°C and 2.2–2.7 times at 275°C for compositions that are modified by carbon nanostructures containing Ni (1:20), Co (1:5), which is due to the release of water and ammonia vapors into the bubbles formed and the pressure in the bubbles increases.

In contrast to the presented picture in modified compositions containing carbon–manganese-containing nanostructures, the heat capacity changes without significant jumps, which is caused by a more gentle flow of gas formation. At the same time, it is possible to note a higher value of the specific heat at a temperature of 375°C of compositions that are modified by carbon nanostructures containing Ni (1:20).

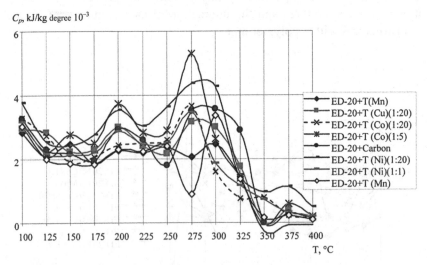

FIGURE 2.38 (See color insert.) Dependence of heat capacity on temperature.

Based on the foregoing, it can be concluded that the most effective compositions for heat capacity are compositions that are modified by carbon nanostructures containing Ni (1:20), Co (1:5).

2.5 CONCLUSIONS

As a result of the research

1. For the first time, it is determined that carbon–metal-containing nanostructures are a new highly effective class of structure-forming foam cells.

2. It has been established that the use of carbon–metal-containing nanostructures lead to an increase in carbon–carbon and carbon–metal groupings in foam boxes.

3. It was established that a synergistic increase in the fire-protective and thermophysical characteristics of intumescent coatings based on the epoxy resin ED-20 modified with ammonium polyphosphate and carbon–metal-containing nanostructures is due to the formation of a stable, ordered foam with small pores resistant to oxidation during the destruction process.

4. It has been found that the joint use of the intumescent coating and the carbonized layer reduces the thickness of the heated layer by thermal shocks with an equi-radiator 1.6 times and 2.7 times compared with the use of the intumescent coating and the carbonized layer, respectively.

5. It has been established that the joint use of the intumescent coating and the carbonized layer contributes to an increase in the fire resistance limit of 1.5–2.0 times compared with the use of only one type of coating.

6. For the first time, it is shown, first, that the heat capacity of compositions modified by structurants is three to seven times higher than the heat capacity of compositions containing only ammonium polyphosphate, which is explained by the different degree of structure formation of the compositions. Second, the heat capacity of the composition containing PFA (Composition 5) sharply increases by a factor of 10–11, at temperatures in the region of 373–403 K and 430–480 K, which is apparently due to the release of water vapor into the bubbles and ammonia and increasing pressure in the bubbles. Unlike the picture presented in the modified compositions, the heat capacity changes without significant jumps, which is due to a more calm flow of the gassing process.

7. It was established for the first time that, with an increase in pyrolysis temperature, there are differences in the chemical structure of the surface and the volume of the intumescent coating, with the help of XPS that graphite-like substances are formed on the internal surface of the bubbles in intumescent coatings. The use of carbon–metal-containing nanostructures containing nickel leads to an almost three-fold increase in carbon-carbon bonds in foam boxes.

8. Using AFM and SEM methods, it has been determined that the use of carbon–metal-containing nanostructures in intumescent coatings makes it possible to obtain a foam box, which is a set of bubbles whose dimensions are approximately the same and relatively small. Inside, the foam box has a porous structure with very thin walls and high air content in the inter-porous space, which leads to the ordering of the structure of the foam boxes.

9. For the first time it is determined that for the intumescent systems containing cobalt, the strength of the foam boxes is the higher, the smaller the interplanar distance corresponding to the graphite-like peak 2, and the higher its intensity. The observed dependence of the foam strength increases with the packing density of the layers in the foam boxes (decrease in the interplanar distance) and the number of sections with such a structure (the growth of the relative intensity of the peak). For intumescent systems containing copper, there is a reverse trend. The decrease in strength of foam boxes is associated with a change in the superstructure of the carbonization product of gels. With an increase in the relative amount of PVA in gels, the MNT structure in the product changes from the intertwined net to direct single MWT. Accordingly, the mechanical strength of the foam boxes containing MNT in the form of a woven "network" exceeds the strength of the foam boxes containing straight single MNTs. In a similar manner, the strength of the foams of nickel-containing compositions decreases. At close values of interplanar distances, maximum strength is not observed for systems with peak intensity. Addition of graphite slightly increases the strength of foam boxes, and activated carbon is reduced. Such a reduction in strength is also due to the superstructure of the crushed graphite and activated carbon present in the system in the form of much larger (millimeters and micrometer sizes) than multi-walled CNTs of particles.

10. The principles of creating fireproof materials containing nanostructures are substantiated, consisting in the fact that nanostructures are stimulators of the formation of nanophases in the
material, which contribute to an increase in the "embryos" of the
coked surface with increasing temperature.

ACKNOWLEDGMENT

This work was carried out with financial support from the Research
Program of the Ural Branch of the Russian Academy of Sciences: the
project 15-10-1-23.
The results of the research were obtained during the implementation of
innovative works under the Federal Target Program "Integration," project
No. A 0014, Б 0074.

KEYWORDS

- ammonium polyphosphate
- atomic force microscopy
- differential thermal analysis
- fire-retardant system
- foam coke
- multilayered carbon nanotubes
- nanotubes
- polymer synthetic materials
- polyvinyl alcohol
- single-walled carbon nanotubes

REFERENCES

1. Kamanina, N. V.; Voronin, Yu. M. Microscopic Studies of the Polyimide-C_{70} System.
Lett. Zh. **2002**, *28*(21), 6–10.

2. Gladchenko, S. V.; Polotskaya, G. A.; Gribanov, A. V.; Zgonnik, V. N.; Gladchenko, S. V. Investigation of Solid-Phase Polystyrene-Fullerene Compositions. *ZhTF* **2002**, *72*(1), 105–109.

3. Baranova, V. A.; et al. A. *Phosphorus-Containing Flame Retardants;* NIITEKhim: M., 1978.

4. Pozdnyakov, O. F.; Redkov, B. P.; Pozdnyakov, A. O. Thermal Stability of Polystyrene Films Chemically Bonded to C60 Fullerene. Thickness Effect in Submicron Range. *Lett. ZhTF* **2002**, *28*(24), 53–57.

5. Dai, H.; et al. *Nature (London)* **1996**, *384,*147.

6. Kuzmany, S. C.; et al. *Fullerenes and Fullerene Nanostructures;* World Scientific: Singapore, 1996; p 250.

7. Badamshina, E. R.; Gafurova, M. P.; Estrin Y. I. *Success Chem.* **2010**, *79*(11), 1027–1063.

8. Du, N.; Bai, J.; Chens, Y.-M. *J. Express Polivmer Lett.* **2007**, *1*(5) 253–273.

9. Xie, H.; Liu, B. *J. Polym. Sci. Part B. Polymer Phys.* **2004**, *42*, 3701–3712.

10. Bogatov, B. A.; Kondrashov, S. V.; Mansurova, I. A.; Minakov, V. T.; Anoshkin, I. V. On the Mechanism of Strengthening Epoxy Resins with Carbon Nanotubes. *All Materials. Encyclopedic Reference Book,* **2012**, *4.*

11. Tarasov, V. A.; Stepanishev, N. A. Creation of Nanocomposites Based on the Introduction of Carbon Nanotubes into Epoxy and Polyester Binders. *Materials of the All-Russian Scientific and Practical Internet Conference with International Participation "High Technologies in Mechanical Engineering,"* Samara, Nov 17–20, 2010.

12. Stepanishev, N. A.; Tarasov, V. A. *Strengthening the Polyester Matrix by Carbon Nanotubes to Accelerate Gelling;* Bulletin of the MSTU. N. E. Bauman. Ser. Instrument Making, 2010; 53–65.

13. Tkachev, A. G.; Kharitonov, A. P.; Simbirtseva, G. V.; Kharitonova, L. N.; Blokhin, A. N.; Dyachkova, T. P.; Druzhinina, V. N.; Maksimkin, A. V.; Chukov D. I.; Cherdyntsev, V. V. Strengthening of Epoxy Materials by Fluorinated Carbon Nanotubes. (URL: http://science-education.ru/ru/article/view?id=12620)

14. Brauman, S. K. Char-forming Synthetic Polymers. Combustion Evaluation. *J. Fire Retard. Chem.* **1979**, *6*(4), 244–265, 266–275.

15. Lipatov, Y. S. *Interphase Phenomena in Polymers;* Science: Dumka, Kiev, 1980; p 259.

16. Shuklin, S. G.; Kodolov, V. I.; Kuznetsov, A. P. Investigation of Physicochemical Processes in Expanding Compositions Based on Epoxy Resin. *Chem. Phys. Mesoscopy* **2000**, *2*(1), 5–11.

17. Didik, A. A.; Kodolov, V. I.; Volkov, A. Y.; Volkova, E. G.; Halmayer, K. Kh. A Low-Temperature Method for Obtaining Carbon Nanotubes in a Condensed Phase. *Inorg. Mater.* **2003**, *39*(6), 693–697.

18. Didik, A. A.; Kodolov, V. I.; Volkov, A. Y.; Volkova, E. G. Synthesis of Metal-Containing Tubulenes from Polyvinyl Alcohol at a Temperature of 250°C. *CPM* **2002**, *4*(2), 214–223.

19. Emmanuel, N. M.; Knorre, L. G. *Course of Chemical Kinetics;* High School: M., 1969, p 431.

20. Bulgakov, V. K.; Kodolov, V. I.; Lipanov, A. M. Modeling of Burning of Polymeric Materials. Chemistry: M., 1990, p 238.

21. Berlin, A. A. Combustion of Polymers and Polymeric Materials of Low Flammability. *Sorovsky Educ. Journal.* **1996**, *9,* 57–63.
22. Antonov, A. V.; Reshetnikov, I. S.; Khalturinskii, N. A. Burning of Coke-Forming Polymer Systems. *Uspekhi Khimii.* **1999**, *68*(7), 663–673.
23. Mashlyakovsky, L. N.; Lykov, A. D.; Repkin, V. Yu. *Organic Coatings of Low Flammability;* Chemistry: L., 1989; p 184.
24. Gusev, A. I. Effects of the Nanocrystalline State in Compact Metals and Compounds. *UFN* **1998**, *168*(1), 55–83.
25. Jain, P. K.; Mahajan, Y. R.; Sundararajan, G. Development of Carbon Nanotubes and Polymer Composites. *Carbon Sci.* **2002**, *3*(3), 142–145.
26. Aseeva, P. M.; Zaikov G. Ye. *Combustion of Polymeric Materials;* Science: M., 1981; p 280.
27. Ubbelohde, A. R.; Lewis, F. A. *Graphite and Its Crystalline Compound;* Mir: M., 1965; p 256.
28. Berber, S.; Kwon, Y.-K.; Tománek, D. Unusually High Thermal Conductivity of Carbon Nanotubes. *Phys. Rev. Lett.* **2000**, *84*(20), 4613–4616.
29. Lykov, A. V. *Theory of Heat Conductivity;* High School: M., 1967.
30. Anufrieva, E. V.; Krakovyak, M. G.; Anan'eva, T. D.; Nekrasova, T. N.; Smyslov, R. Y. Interaction of Polymers with C_{60} Fullerene. *FTT* **2002**, *44*(3), 443–444.
31. Lau, K.-T.; Hui D. Effectiveness of Using Carbon Nanotubes as Nano-Reinforcements for Advanced Composite Structures. *Carbon* **2002**, *40*(9), 1605–1606.
32. Morrison, S. *Chemical Physics of the Surface of a Solid Body;* Mir: Moscow, 1980; p 488.
33. Bukharaev, A. A. Surface Diagnostics with the Help of Scanning Tunneling Microscopy. *Fact. Lab.* **1994**, *60*(10), 15–25.
34. Maganov, S. N.; Whangbo, M.-H. *Surface Analysis with STM and AFM;* VCH: Weinheim, New York, Basel, Cambridge, Tokyo, 1996; p 50.
35. Sinniah, S. K.; Steel, A. B.; Miller C. J. Solvent Exclusion and Chemical Contrast in Scanning Force Microscopy. *J. Am. Chem. Soc.* **1996**, *118,* 8925–8931.
36. Berman, R. *Thermal Conductivity of Solids;* Mir: M., 1979; p 170.
37. Shashkov, A. G.; Tyukaev, V. I. *Thermophysical Properties of Decomposing Materials at High Temperatures;* Science and Technology: Minsk, 1975.38. Povstukhar, V. I.; Bystrov, S. G.; Mikhailova, S. S. Study of the Local Chemical Structure of the Surface by Means of Probe Methods, Materials Vseros. Sovshch. "Probe microscopy—99". Nizhny Novgorod, 1999, 305–309.

CHAPTER 3

USE OF GENETIC ALGORITHMS TO SOLVE OPTIMIZATION PROBLEMS: A CRITICAL OVERVIEW AND A VISION FOR THE FUTURE

SUKANCHAN PALIT[1,2]

[1]Department of Chemical Engineering, University of Petroleum and Energy Studies, Bidholi via Premnagar, Dehradun, Uttarakhand 248007, India

[2]43, Judges Bagan, Haridevpur, Kolkata 700082, India

E-mail: id-sukanchan68@gmail.com; sukanchan92@gmail.com.

ABSTRACT

Human civilization and human scientific endeavor are today on the path toward immense scientific regeneration and scientific rejuvenation. Today, the world of engineering science is witnessing drastic challenges. Applied mathematical tools are the revolutionary branches of scientific research pursuit today. In a similar vein, genetic algorithms (GAs) are today challenging the scientific landscape. Chemical process engineering, petroleum engineering science, and environmental engineering are the avenues of scientific endeavor which has applications of GA and multi-objective optimization. In this chapter, the author repeatedly stresses upon the scientific success and the deep scientific potential of GA in the avenues of research pursuit in petroleum engineering, petroleum refining, and chemical process engineering. Scientific validation and technological vision are the challenges of present-day human civilization. Humankind's immense scientific intellect, the futuristic vision of applied mathematics, and the intellectual prowess of scientific rigor will all lead a long and visionary way in the true realization of GA and multi-objective optimization (MOO) today. There is an immense importance of GA in designing of petroleum refining units. This area is deeply investigated in this chapter. The author also pointedly focuses on the vast importance of GA and optimization in chemical engineering and environmental engineering systems.

Scientific imagination and scientific genesis are of utmost importance on the path toward true realization of effective modeling, simulation, control, and optimization of chemical engineering and petroleum engineering systems. The status of research forays in GA and optimization science are far-reaching and replete with cross-boundary endeavor. The need of human society is energy sustainability and energy security. This chapter opens up new windows of scientific instinct and scientific innovation in GA and evolutionary computation.

3.1 INTRODUCTION

The vast and versatile world of genetic algorithm (GA) is ushering in a new era of scientific regeneration and deep scientific vision. Today, the scientific world stands in the midst of deep comprehension and definitive vision. The application of GA in design of chemical and petroleum engineering systems are replete with vision and scientific forbearance. Scientific genre and scientific genesis are in the state of immense revival. Technological advancements and scientific validation are rapidly changing the scientific landscape of human civilization and human scientific endeavor. Multi-objective optimization (MOO) is the need of the hour in the scientific truth in designing chemical and petroleum engineering systems. Today, GA is a computational tool which is surpassing vast and versatile scientific frontiers. Petroleum engineering systems and chemical engineering systems and their design are the challenging research questions of today. The main vision and purpose of this chapter targets the scientific doctrine and fundamentals of use of GAs in solving optimization problems. Scientific vision and deep scientific fortitude are the necessities of scientific and academic rigor today.

3.2 AIM AND OBJECTIVE OF THIS STUDY

Today, GA and MOO are on the path toward immense scientific introspection and deep scientific forbearance. The world of technology and science is moving at a rapid pace breaking visionary scientific boundaries. Applied mathematics and chemical process engineering are two opposite sides of the visionary coin. The main aim and objective of this study targets the scientific intricacies and scientific hindrances in the application of GAs in solving optimization problems. MOO is widely used in the design and optimization of chemical engineering and petroleum engineering systems. The challenge and the vision of science veritably go beyond scientific imagination and scientific truth. This chapter elucidates on the scientific success and the deep scientific comprehension in the application of GA in solving optimization problems. Humankind is today in a state of immense scientific rejuvenation. The march of science and applied mathematics, the immense scientific prowess of human civilization, and the futuristic vision of technology will all lead a long and visionary way in the true realization of optimization science. Optimization science is the fountainhead of scientific progress in applied mathematics, petroleum engineering, and chemical process engineering. Modeling, simulation, optimization, and control are the scientific pillars of true emancipation of petroleum engineering science and chemical engineering science. This study goes beyond scientific imagination and scientific forbearance and presents before the reader the deep scientific cognizance and sagacity in the research pursuit in MOO and GA with a clear vision toward furtherance of science, engineering, and applied mathematics.[1,2]

3.3 LITERATURE REVIEW

Scientific doctrine and scientific cognizance are today in a state of deep comprehension as science, engineering, and human civilization march forward toward a newer visionary realm. GA, MOO, and applied mathematics are changing the scientific landscape of engineering science emancipation and technological advancement. Humankind's immense scientific prowess, the vast futuristic vision, and the veritable needs of human society are the forerunners toward a newer visionary era in scientific emancipation. Petroleum engineering science today is in a state of immense scientific quagmire. The vexing issue of depletion of fossil-fuel resources is challenging the scientific domain today. Petroleum refining and chemical process design need to be reenvisioned and revamped with the passage of scientific history, scientific vision, and time. This chapter widely presents the scientific success, scientific forbearance, and the scientific candor of the intricacies in GA application in the design of petroleum engineering and chemical engineering systems. This literature review revamps and redefines the immense scientific importance and the deep technological vision in the GA applications in modeling, simulation, and optimization of petroleum engineering and chemical engineering systems. Rangaiah[3] discussed lucidly in a watershed treatise the scientific success of multi-objective applications in process engineering systems. GAs, according to the author, have revolutionized the field of water system engineering and bioengineering. Their chapter is a veritable proponent of evolutionary computation. Deb[4] propounded the scientific success of MOO and GA. Systematically, this chapter challenged the scientific frontiers of GA. This is a well-researched chapter which revolutionized the paradigm of GA. Bandyopadhyay et al.[5] dealt lucidly on a simulated annealing (SA)-based MOO. This is another scientific frontier of evolutionary computation. Ramteke et al.[6] challenged the veritable scientific fabric of MOO. The authors focused on the reactor simulation of polymerization processes. Chemical process engineering is veritably the challenging arenas of GA and MOO. The vision and the mission of this chapter go beyond scientific imagination and deep profundity.

3.4   WHAT IS GENETIC ALGORITHM (GA)?

GA is a visionary path toward scientific endeavor. Today, human scientific forays, the vast scientific far-sightedness, and the scientific insight are all leading a long and visionary way in the true emancipation of petroleum engineering and chemical process engineering. The GA is a method to solve both constrained and unconstrained optimization problems that are based on natural selection and the technique that drives biological evolution. This tool is a hallmark of applied mathematics and applied science. The GA repeatedly modifies a population of individual solutions. GA has diverse applications today as science and engineering enter into a visionary realm. In applied computer science and operations research, a GA is a metaheuristic inspired by the process of natural selection that largely belongs to the larger class of evolutionary algorithms. GAs are commonly applied to generate high-quality solutions in optimization and search problems.

3.5   WHAT IS MULTI-OBJECTIVE OPTIMIZATION (MOO)?

Today, MOO and optimization of petroleum refining units are the two opposite sides of the visionary coin. Design of chemical engineering and petroleum engineering systems are the scientific imperatives and the prerequisites toward the furtherance of science and engineering. Optimization science is ushering in a new era in the field of petroleum refining, chemical process engineering, and applied mathematics. The challenge of science needs to be revisited with every step of scientific research pursuit. MOO is an area of multi-criteria decision-making (MCDM) that is concerned with mathematical optimization problems involving more than one objective function to be optimized simultaneously. Technology and engineering science are today moving far ahead in present-day human civilization. Research questions in optimization science need to be reenvisioned and reenvisaged as science crosses one visionary paradigm over another. This scientific vision needs to be redefined and reframed as applied mathematics, optimization, and engineering science move forward.[1,2]

3.6   SCIENTIFIC DOCTRINE OF PETROLEUM ENGINEERING AND CHEMICAL PROCESS ENGINEERING

Human civilization and human scientific endeavor are the forerunners of a new scientific order and new scientific emancipation. Technology of petroleum engineering and chemical process engineering are changing the scientific frontiers and moving toward a newer scientific evolution. Petroleum engineering systems such as petroleum refinery today stands in the midst of deep crisis and deep scientific introspection. Each petroleum refining unit is of vital importance in the process of energy sustainability and the vast world of energy engineering. Depletion of fossil-fuel resources is challenging the scientific landscape and scientific mind-set of future generations. The vision of science is retrogressive and needs to be deeply restructured. The vision and doctrine of science are challenging and groundbreaking in the field of modeling, simulation, optimization, and control of petroleum refining units. This chapter opens up the world of scientific imagination and scientific profundity in these areas of scientific forays, particularly, design of petroleum refining units and chemical engineering systems. Human scientific endeavor in sustainability science and petroleum engineering science today stands in the midst of deep scientific rejuvenation and fortitude. The author deeply ponders upon the entire gamut of scientific success and scientific progeny in GA and multi-objective applications with the true vision of furtherance of science and engineering.[1,2]

3.7   SCIENTIFIC TRUTH AND SCIENTIFIC VISION OF ENERGY SUSTAINABILITY

Energy sustainability and petroleum engineering science are today two opposite sides of the visionary coin. The vast scientific truth, the scientific genesis, and the scientific fortitude will all lead a long and visionary way in the true realization of energy sustainability. Energy, water, food, and environment are the utmost needs of the human civilization and human scientific endeavor. Depletion of fossil-fuel resources and frequent environmental catastrophes are the grave concerns for human civilization. Environmental sustainability also needs to be reenvisioned and restructured as human civilization moves from one visionary paradigm toward another. Today, scientific truth and scientific vision in the domain of energy and environmental sustainability are gaining immense heights with the passage of scientific history and the visionary time frame. Scientific validation and technological advancements are the necessities of the deep scientific endeavor in sustainability science. The vision of Dr. Gro Harlem Brundtland, former Prime Minister of Norway, and the proponent of sustainability needs to be reenvisioned and reenvisaged. The major areas of research pursuit in petroleum engineering science are the design of petroleum refining units.

At its most general level, sustainability deeply refers to the capacity to continue an activity or process indefinitely. This visionary terminology goes beyond scientific imagination and scientific justification. Environmental and energy sustainability are today on the path toward newer scientific regeneration and are the utmost need of the hour. Petroleum refining and the scientific strides of petroleum engineering science are the imperatives of successful sustainability. Mathematical tools such as GA and optimization and its vast and varied applications are ushering in a new era in science and engineering, particularly, petroleum engineering science and chemical process engineering.

3.8 RECENT SCIENTIFIC RESEARCH PURSUIT IN GA AND MOO

The vision and the challenge of science are surpassing vast and versatile scientific frontiers. GA and optimization science are the two opposite sides of the visionary coin. Petroleum engineering science and chemical process engineering are the two pillars of mathematical applications such as GA, optimization, and evolutionary computation. Technology and engineering science are highly challenged today with the passage of scientific history and time. Recent scientific research pursuit in GA needs to be redefined and restructured. Today, petroleum refining is a vital avenue of scientific research pursuit. GA applications in the optimization of a petroleum refining unit are opening up new ventures and new ideas and scientific instinct.

Rangaiah[3] discussed with immense lucidity and cogent insight MOO with techniques and applications in chemical engineering. The scientific success, the deep scientific vision, and the world of challenges are today ushering in a newer eon in the field of applied mathematics and chemical process technology. Human civilization and human scientific endeavor are in the process of immense scientific rejuvenation. Optimization is immensely essential for reducing material and energy requisites as well as the harmful environmental effects of chemical processes. Optimization leads to better design and operation of chemical processes as well as sustainable techniques. Many applications of optimization involve several objectives, some of which are extremely conflicting.[3] MOO is veritably required to solve the resulting problems in these applications. This is the reason why MOO has attracted immense scientific interest.[3] Today, MOO has vast and varied applications and veritably crossing scientific frontiers. Technological vision, scientific genesis, and deep scientific cognizance will all lead a long and visionary way in the true realization of applied mathematical tools and chemical process engineering.[3] Rangaiah,[3] in the well-researched treatise, elegantly ponders upon the immense scientific far-sightedness, the scientific profundity, and the deep scientific doctrine in MOO applications of process systems engineering. The well-researched book provided an introduction to MOO with a vastly realistic application, namely the alkylation process optimization for two objectives.[3] The other vision of the book is reviewing nearly 100 chemical engineering applications of MOO since the year 2000 to mid-2007.[3] The challenge and vision of technology, the deep futuristic vision, and the success of sustainability science are the veritable forerunners toward a newer scientific vision of applied mathematics tools such as optimization and GA. The other vast visions of the treatise are review of multi-objective evolutionary algorithms (MOEAs) in the context of chemical engineering, multi-objective GA and SA as well as their jumping gene adaptations, surrogate-assisted MOEA, interactive MOO in process design, and two methods for ranking the Pareto solutions.[3] Human scientific research pursuit in MOO and GA are opening new windows of deep scientific vision and scientific forbearance in the area of energy sustainability and the holistic domain of applied science and applied mathematics.[3] The vast volume of research expertise challenges the new scientific order in evolutionary computation and also the vast domain of GA, MOO, multi-objective SA, and differential evolution.[3] The scientific forays in chemical process engineering and petroleum engineering science are challenging the vast scientific landscape and the deep scientific vision of engineering science.[3] The treatise vastly comprehended the challenge, the success, and the mission behind chemical engineering aspects of MOO.[3]

Rangaiah[7] dealt lucidly on the topic of process optimization and its interfaces with MOO. The vast scientific success, the scientific candor, and the scientific expertise in the domain of MOO are deeply dealt with immense foresight as science and engineering moves toward a newer visionary era.[7] Optimization deeply refers to finding the value of decision variables which correspond to and provide the maximum and minimum of one or more desired objectives. The visionary splendor of science, the challenge, and vision of technology and the futuristic vision of scientific discernment are today the forerunners of scientific forays in process optimization. Optimization veritably finds many applications in engineering, science, business, economics, and so forth, except that, in these applications, quantitative models and definite methods are employed.[7] The vast vision of applied science, applied mathematics, and process optimization are highly challenged as human civilization and human scientific endeavor enters into a newer visionary realm. The challenge of process optimization needs to be redefined as science and technology move forward. Optimization has many applications in chemical, mineral processing, oil and gas, petroleum, pharmaceuticals, and related manufacturing industries.[7] Process modeling and optimization along with control deeply envisage and re-envision the area of process systems engineering with a deep vision toward furtherance of science and engineering. The main focus and the deep vision of the optimization of chemical processes so far have been optimization of one objective at a time (i.e., single-objective optimization, SOO).[7] However, practical scientific forays involve several objectives to be deeply considered simultaneously. These vast objectives can include capital cost/investment, operating cost, profit, payback period, selectivity, quality and recovery of product, conversion, energy required, efficiency, process safety, operation, robustness, and so forth. Scientific challenges in process optimization will open new degrees of innovation and new windows of scientific instinct in decades to come. This challenging arena of process optimization is of utmost need in the scientific struggle and vision toward engineering emancipation. Technology and engineering science are highly advanced today and breaking visionary barriers. Today, MOO stands in the midst of vast scientific vision and deep scientific fortitude.[7] MOO, also known as multi-criteria optimization, particularly outside engineering, refers to finding values of decision variables which correspond to and provide the optimum of more than one objective. The technique is mathematically sound and robust.[7] Unlike in SOO which gives an unique solution, there will be many optimal solutions for a multi-objective problem; the exception is when the objectives are not conflicting in which case only one unique solution is expected. This treatise gives a vast and rigorous approach toward multi-criteria optimization with the sole vision of furtherance of science and engineering.[7] The relevance and importance of MOO in chemical engineering is vastly advancing, which has been partly motivated by the availability of new and effective methods for solving multi-objective problems as well as increased computational resources. Scientific vision, scientific profundity, and deep scientific revelation are the imperatives of research pursuit today. Applied mathematics and optimization science are on the threshold of newer techniques and visionary paradigm. This treatise targets these research questions. In general, an MOO problem will have two or more objectives involving many decision variables and constraints.[7] The success of optimization is well researched and technologically highly advanced. Today, scientific vision and scientific fortitude are in a state of immense catastrophe with the ever-growing concerns of depletion of fossil-fuel resources and frequent environmental disasters.[7] Here comes the need of applied sciences application. In this treatise, the author repeatedly stresses upon the vast intricacies in MOO and application of GA in process systems engineering with the sole vision of advancement and true realization of engineering science.[7]

Masuduzzaman et al.[8] discussed with lucid and cogent details of MOO applications in chemical engineering. The chapter reviewed process design and operation, biotechnology and food industry, petroleum refining and petrochemicals, pharmaceuticals and other products/processes, and polymerization. Scientific sagacity, scientific revelation, and deep scientific vision are the hallmarks of this chapter.[8] Optimization refers to obtaining the values of decision variables, which correspond to the maximum or minimum of one or more objective functions.[8] Major areas of research in optimization and its applications in chemical engineering consider only one objective function, probably due to the available computational resources including methods.[8] However, most real-world chemical engineering research questions involve one or more objectives which are mainly computational in nature. The vision of science, the challenge of technology, and the scientific empowerment will all lead a long and visionary way in the true emancipation of mathematical tools in design today. In this chapter, the authors summarized MOO problems in all areas of chemical engineering reported in journal publications from 2000 until mid-2007.[8] Technological vision, deep scientific motivation, and the futuristic vision are the torchbearers toward a newer eon in the field of chemical process engineering and petroleum engineering. Every detail was recorded and it also included all reported MOO applications of chemical engineering in this chapter.[8] The authors repeatedly stressed upon the vast outcome and scientific results of MOO applications in chemical engineering. Process design and operation, which plays a pivotal role in chemical engineering, have attracted many applications in MOO since the year 2000.[8] Human scientific aura, scientific vision, and deep scientific profundity are the torchbearers toward a newer era in chemical process engineering and MOO. In all, there are 35 applications of MOO for process design and operation. These encompasses fluidized bed dryer, cyclone separator, a pilot-scale venturi scrubber, hydrogen cyanide production, heat exchanger network, grinding, froth flotation circuits, simulated moving bed (SMB) and related separation systems, thermal and pressure swing adsorption, toluene recovery, heat recovery system, a cogeneration plant, batch plants, safety-related decision-making, chemical process safety, system reliability, proportional–integral controller design, waste incineration, parameter estimation, industrial ecosystems, scheduling, and supply chain networks, and so forth.[8] The treatise is a comprehensive review of the vast and versatile world of chemical process engineering and MOO.[8] Technology and engineering science are today in a state of deep scientific catastrophe. Human scientific regeneration today stands in the midst of deep scientific comprehension. Applied science and applied mathematics are the forerunners toward a newer visionary era in optimization science. In the similar vein, GA and scientific profundity needs to be reenvisioned and reenvisaged with every step of human scientific research pursuit.

Jaimes et al.[9] deeply comprehended MOEAs in a review of the state-of-the-art and some of their applications in chemical engineering. The status of research pursuit in evolutionary algorithms is breaking and surpassing visionary frontiers. The challenge, the vision, and the targets of human scientific endeavor are changing veritably the scientific fabric of human scientific research pursuit in MOO and GA.[9] In the treatise, the authors repeatedly stressed upon the scientific success, the deep scientific genesis, and the sound scientific progeny in the pursuit of science in GA. The authors provided a general overview of evolutionary MOO with particular emphasis on algorithms in current applications. Technological advancements, scientific motivation, and the candor of engineering science are today revolutionizing the state of human endeavor.[9] The challenge and vision of MOO applications are immense and far-reaching. The treatise definitely opens up new windows of scientific empowerment in MOO and GA in decades to come. In this treatise, several applications of these algorithms in chemical engineering are deeply discussed and analyzed.[9] The deep structure of scientific research pursuit in MOO needs to be reenvisioned and reenvisaged with the passage of scientific history and visionary time frame.[9] The solution of problems having two or more (normally conflicting) objectives has become exceedingly common in the last few years in diverse areas of engineering science. Such problems are called multi-objective and need to be reframed with the passage of scientific vision. These mathematical problems can be solved using either mathematical programming techniques or using metaheuristics.[9] In either case, the visionary concept of Pareto optimality is normally adopted. Of the many metaheuristics available, evolutionary algorithms have become increasingly popular because of their ease of implementation and high effectiveness.[9] The robustness of MOO technology, the vast scientific prowess, and the visionary ideas will lead a long and visionary way in the true emancipation of optimization science. The first MOEAs were introduced in the scientific arena in the 1980s but they became popular only in the mid-1990s.[9] Human scientific regeneration and technological vision are today in a state of success and robustness. MOO and GAs are the scientific masterpieces of present-day human civilization. This is a watershed text which throws immense light on the scientific success, the scientific forbearance, and the vast challenges in the field of MOO applications. The structure of this research foray is highly far-reaching and needs to be restructured as human civilization and human scientific genre gains immense height with the passage of history and visionary time frame.[9] The chapter provided a short introduction to MOEAs, presented and lucidly dealt with historical perspective. It definitely reviewed some of the most representative work regarding their use in chemical engineering forays. Finally, it provided a brief discussion of some of the main Internet resources currently available for interested researchers. Some well-known application areas dealt with are optimization of industrial nylon 6 semi-batch reactor, optimization of industrial ethylene reactor, optimization of industrial styrene reactor, optimization of an industrial hydrocracking unit, optimization of semi-batch reactive crystallization process, optimization of SMB process, biological and bioinformatics problems, optimization of a waste incineration plant, chemical process system modeling, and a whole gamut of innovative ideas and research forays.[9] This treatise widely revisits the challenges, the deep vision, and the innovative success areas on the path toward the true emancipation of optimization science.[9]

Ramteke et al.[10] discussed with lucid and cogent insight multi-objective GA and SA with the jumping gene adaptations. Two popular evolutionary tools are used for MOO problems, namely GA and SA which are discussed with deep foresight. These tools are inherently more robust than conventional optimization techniques.[10] The vast scientific success, the technological revolution in MOO applications, and the deep futuristic vision will all lead a long and challenging way in the true realization of engineering science and the holistic world of optimization science. In the chapter, detailed descriptions of GA and SA with the various jumping gene adaptations are deeply presented and three benchmark problems are solved using them.[10] Most real engineering problems require the simultaneous optimization of several objectives (MOO) that cannot be compared easily with one another, that is, are noncommensurable.[10] Technological vision and deep scientific challenges are the visionary forerunners of engineering science of optimization and GA.[10] In the well-researched treatise, the authors delve deep into the unknown world of human research pursuit in applied mathematics and optimization science. Very popular and robust techniques such as GA and SA are vastly used to solve these intricate MOO problems.[10]

Miettinen et al.[11] discussed with deep foresight the challenges in the use of interactive MOO in chemical process design. The path toward visionary ideas in MOO needs to be reenvisaged and reenvisioned today.[11] Problems in chemical process engineering, like most real-world optimization problems, typically, have some conflicting performance criteria or objectives and they often are computationally demanding, which sets special requirements on the optimization methods used.[11] In the chapter, the authors pointedly focuses on some shortcomings of some widely and basic methods of MOO. Technological profundity, the deep scientific revelation, and the vast scientific success are the imperatives of scientific endeavor today.[11] The authors discussed usefulness of interactive approaches in chemical process design by summarizing some visionary studies related to, for example, papermaking and sugar industries. Problems involving multiple conflicting criteria or objectives are generally known as MCDM problems. In such problems, instead of well-defined single optimal solutions, there are many compromise solutions, so-called Pareto optimal solutions that are mathematically incomparable.[11] In the MCDM literature survey, solving an MOO problem is usually understood as helping a human decision maker in considering the multiple objectives simultaneously and in finding a Pareto solution that pleases him/her most. The challenge of engineering science and applied mathematics are slowly opening up new vistas of research avenues.[11]

Garg[12] discussed with deep and cogent insight array informatics using multi-objective GAs with special emphasis on gene expressions and gene networks. Multi-objective GAs are the revolutionary avenues of research endeavor today. The futuristic vision, the deep scientific cognizance, and the vast scientific vision to move forward will all lead a long and visionary way in the true emancipation and true realization of optimization science today.[12] In this phenomenal work, development of a robust clustering algorithm and a graph—theoretical model for reverse engineering the gene networks and their implementation using nondominated sorting Genetic Algorithm-II (NSGA-II) are reported. Clustering results on synthetic datasets as well as on a real-life dataset are extremely encouraging. One of the most promising areas of computational biology is the task of deciphering underlying regulatory genetic networks mathematically from expression data resulting from large-scale arraying techniques. Computational biology is the forerunner of diverse areas of scientific research in GA.[12] Human scientific endeavor in computational biology is highly advanced today. Technology has few questions to diverse intricacies in computational biology. The author in the treatise merged biology with optimization science. Cells are the structural and functional units of all living organisms. The traditional techniques of molecular biology were based on one gene–one experiment and the output was very limited. It was extremely difficult to discern about the various interactions.[12] There was always a need for monitoring number of genes simultaneously.[12] These concepts resulted in the development of microarray technology. Scientific vision and scientific forbearance are the pillars of modern computational biology.[12] This microarray technology, developed in the past few decades, makes it possible for the researchers to have a better picture of interaction among thousands of genes simultaneously, by monitoring the whole genome on a single chip.[12] The author in the treatise touched upon preprocessing of microarray data, gene expression profiling, and gene network analysis and network analysis. There is a vast interface of optimization and biological sciences in the well-researched treatise.[12]

Andersson[13] deeply elucidated in a research thesis MOO in engineering design with vital emphasis on fluid power systems applications.[13] Technology and engineering science of MOO are veritably entering into a new era. The challenge and the vision of scientific discernment in optimization are also entering into a new era.[13] The thesis focused on how to improve design and development of complex engineering systems by the introduction of simulation and optimization tools. Scientific aura, scientific candor, and the deep scientific profundity are veritably changing the scientific landscape of optimization science.[13] This treatise is a phenomenal presentation of optimization science. Engineering design problems often consists of many conflicting objectives. In many application areas, the multiple objectives are aggregated into one single objective function. The march of engineering science and applied mathematics are evolving into a newer scientific regeneration.[13] In this treatise, a method is presented in which the Design Structure Matrix and the Relationship Matrix from the House of Quality method are vastly applied to support the formulation of the objective matrix.[13] The thesis elucidated upon the engineering design process, optimization in engineering design, optimization methods, applications, and robustness versus optimality. The first aim of the thesis was to develop a framework where optimization is employed in order to speed up and improve the design of complex systems based on simulations.[13]

Swan[14] discussed with deep and cogent insight in his doctoral thesis optimization of water treatment works using Monte Carlo methods and GAs. Optimization of potable water treatment could result in substantial cost savings for water companies and their customers.[14] In order to reenvision this issue, computational modeling of water treatment works using static and dynamic models was deeply investigated alongside the application of optimization techniques including GAs and operational zone identification. Technology and engineering science are ushering in a new era in the arena of scientific emancipation.[14] The scientific success, the deep scientific vision, and the vast scientific forbearance will all lead a long and visionary way in the true realization of optimization science and water treatment. In the thesis, these optimization methods were explored with the assistance of case study data from an operational work.[14] Scientific enshrinement is at its pinnacle as science and engineering moves forward. Human civilization and deep human scientific endeavor are in a state of immense introspection.[14] Technology needs to be rebuilt and science reenvisioned with the progress of humankind. It was found deeply that dynamic models were more accurate than static models at predicting water quality of an operational site but that the root mean square of the models was within 5% of each other for key performances. Human scientific research pursuit today stands in the midst of comprehension and scientific resurgence. Technological vision and scientific pragmatism are the hallmarks of applied mathematics applications. This treatise vastly pronounces the deep scientific comprehension and vast scientific adjudication in the quest toward emancipation of optimization science.[14]

Dawood et al.[15] deeply discussed modeling and optimization of new flocculent dosage and pH for flocculation for the removal of pollutants from wastewater. Technology and engineering science of optimization are ushering in a new era in the field of chemical engineering and environmental engineering. In the paper, a new ferric chloride hybrid copolymer was successfully synthesized by free radical polymerization in solution using ceric ammonium nitrate as redox initiator.[15] The challenge and the vision of engineering science are surpassing scientific imagination and vast scientific boundaries. In the paper, the hybrid copolymer was characterized by Fourier-transform infrared spectroscopy and scanning electron microscopy.[15] Response surface methodology, involving central composite design matrix with two of the most important operating variables in the flocculation process: hybrid copolymer dosage and pH, was used in this study for the optimization of industrial wastewater treatment process.[15] Scientific motivation, deep technological vision, and the vast scientific candor are the hallmarks of research pursuit.[15] Science is a huge unimaginable colossus with a deep and definite vision of its own. This treatise ponders upon the vision, the challenges, and the genesis of the optimization applications in industrial wastewater treatment paradigm.[15]

Shau et al.[16] described with deep lucidity GAs for design of pipe network systems. The vast technological vision and the scientific profundity are the imperatives and pillars of scientific and engineering paradigm. In the last three decades, a significant number of methods for optimal design of pipe network systems have been reenvisioned using linear programming, nonlinear programming, dynamic programming, enumeration techniques, and GA.[16] Applied mathematics, chemical process engineering, and pipeline engineering are the hallmarks of scientific regeneration. The paper explored a GA approach to the design of a pipe network system. The objectives considered are minimization of the network cost.[16] The authors in the paper investigated a case study of Ruey-Fang district water supply system in the research attempt of optimization science.[16] Human scientific vision, scientific aura, and deep scientific candor are the intricate challenges of science. The authors explored the case study and by comparing the data gathered from the case study, the study also aimed to the verification of the efficacy of the process. The challenges and the targets of science are ushering in new eons and newer scientific pragmatism in the arena of optimization science. The deep success, the vast vision, and the scientific profundity are elucidated in details in the study.[16]

Sawaragi et al.[17] deeply discussed with cogent insight the phenomenal work on MOO. The technology and science of optimization are dealt with immense foresight in their book. The authors touched upon mathematical preliminaries, solution concepts, and some properties of solutions, stability, Lagrange duality, conjugate duality, and vast methodology of optimization science.[17] The book presented in a comprehensive way some salient theoretical areas of MOO. Applied mathematics today stands in the midst of introspection and vision. Optimization science is a part and parcel of the human life. The authors touched upon various kinds of decision-making problems with a serious approach toward furtherance of science and engineering.[17,5,4,6]

Technology and engineering science are on the path toward newer glory and intellectual prowess. GA applications in optimization are revolutionizing the scientific landscape. The vast vision, the scientific prowess, and the futuristic vision of GA applications in water treatment and industrial wastewater treatment are veritably ushering in a new eon in science and engineering. In this chapter, the author repeatedly presents and proclaims the engineering and scientific vision of GA applications in diverse areas of research pursuit.

3.9 SCIENTIFIC SAGACITY, SCIENTIFIC FORTITUDE, AND SCIENTIFIC PROFUNDITY IN GA APPLICATIONS

Human scientific research pursuit is today in a state of immense scientific regeneration and deep rejuvenation. Moreover, scientific sagacity, scientific fortitude, and scientific profundity are also in the state of deep vision. GA applications and its vast interface with optimization science are veritably changing the scientific landscape. Human civilization and human scientific endeavor are ushering in a new era in the field of scientific regeneration and introspection.[5,4,6] Science and engineering are huge colossi with a definite and vast vision of their own. Energy and environmental sustainability in a similar vision needs to be revamped and reenvisaged with the passage of scientific history and time. In this treatise, the author repeatedly stresses upon the vast scientific success, the scientific genesis, and the deep introspection behind application of mathematical tools to modeling, simulation, optimization, and control of chemical engineering and petroleum engineering systems. Human scientific research pursuit is at a difficult stage with the march of human civilization. Holistic sustainable development in energy and environment are the other sides of the visionary coin.[5,4,6]

3.10 FUTURE RESEARCH TRENDS AND FUTURE SCIENTIFIC FRONTIERS

Future research trends in the field of GA and optimization science are vast, versatile, and far-reaching. Human society and human scientific research pursuit are today treading a visionary path toward greater emancipation and true realization of energy and environmental sustainability. Holistic sustainable development is the utmost need of the hour. Petroleum engineering science and chemical process engineering are the scientific frontiers of human civilization today. The challenge and the vision of engineering science are changing the path of human scientific regeneration. Future research trends should be targeted toward newer mathematical tools such as evolutionary computation which encompasses GA and SA. MOO and multi-objective SA are changing the face of scientific research pursuit. These two areas of scientific research pursuit need to be explored and deeply investigated with vision and scientific forbearance. Humankind and human scientific research pursuit stands in the midst of deep scientific pragmatism and vast introspection. Future research trends should also target the needs of human society such as energy, water, food, and environment. Future vision and future flow of thoughts should be vastly streamlined toward mathematical techniques in modeling, simulation, optimization, and control of chemical engineering and petroleum engineering systems.

3.11 CONCLUSION AND SCIENTIFIC PERSPECTIVES

Scientific research pursuit in the present-day human civilization stands in the midst of deep scientific vision, fortitude, and scientific introspection. Technology and engineering science have few answers toward the global catastrophes in environmental sustainability, energy sustainability, and energy security. The march of science and engineering today has an unsevered umbilical cord with the progress of applied science and mathematical techniques. Scientific perspectives in GA applications and optimization science are far-reaching and crossing vast and versatile visionary boundaries. Depletion of fossil-fuel resources and environmental catastrophes are today challenging the scientific domain and thwarting the veritable causes of sustainable development. Energy, food, and water are the veritable needs of the human regeneration. This treatise pointedly focuses on the vast vision behind GA applications in optimization of chemical engineering and petroleum engineering systems. Human scientific endeavor is veritably replete with scientific vision and deep introspection. The author with immense foresight treads a visionary path toward the true realization of GA and its vast and varied applications in optimization. Scientific perspectives are in a state of immense regeneration and deep rejuvenation. The world of challenges in GA, the vast futuristic vision, and the success of energy sustainability will all lead a vast and visionary way in the true emancipation of design and modeling of chemical engineering and petroleum engineering systems. The far-sightedness of this chapter goes beyond human scientific imagination and truly reveals the deep scientific genesis behind GA and other mathematical techniques.

ACKNOWLEDGMENT

The author deeply acknowledges with immense respect the contributions of his late father, Shri Subimal Palit, an eminent textile engineer from India who taught the author the rudiments of chemical engineering.

KEYWORDS

- **optimization**
- **genetic**
- **vision**
- **chemical**
- **engineering**

REFERENCES

1. www.google.com (accessed date October 1, 2017)
2. www.wikipedia.com (accessed date October 1, 2017)
3. Rangaiah, G. P. *Multi-Objective Optimization: Techniques and Applications in Chemical Engineering, Advances in Process Systems Engineering;* Rangaiah, G. P., Ed.; World Scientific Publishing Co. Pte. Ltd.: Singapore; 2009; Vol. 1.
4. Deb, K. Multi-Objective Optimization using Evolutionary Algorithms: An Introduction. Kanpur Genetic Algorithm Report No.: 2011003, 2011.
5. Bandyopadhyay, S.; Saha, S.; Maulik, U.; Deb, K. A Simulated Annealing-Based Multiobjective Optimization Algorithm: AMOSA. *IEEE Trans. Evol. Comput.* **2008**, *12*(3), 269–283.
6. Ramteke, M.; Gupta, S. K. Kinetic Modeling and Reactor Simulation and Optimization of Industrial Important Polymerization Processes: A Perspective. *Int. J. Chem. Reactor Eng.* **2011**, *9*.

7. Rangaiah, G. P. Introduction. In *Multi-Objective Optimization-Techniques and Applications in Chemical Engineering, Advances in Process Systems Engineering;* Rangaiah, G. P., Ed.; World Scientific Publishing Co. Pte. Ltd.: Singapore; 2009; Vol. 1 pp 1–25.

8. Masuduzzaman, Rangaiah, G. P. Multi-Objective Optimization Applications in Chemical Engineering. In *Multi-Objective Optimization-Techniques and Applications in Chemical Engineering, Advances in Process Systems Engineering;* Rangaiah, G. P., Ed.; World Scientific Publishing Co. Pte. Ltd.: Singapore; 2009; Vol. 1, pp 27–52.

9. Jaimes, J. L.; Coello Coello, C. A. Multi-Objective Evolutionary Algorithms: A Review of the State of the Art and Some of their Applications in Chemical Engineering. In *Multi-Objective Optimization: Techniques and Applications in Chemical Engineering, Advances in Process Systems Engineering;* Rangaiah, G. P., Ed.; World Scientific Publishing Co. Pte. Ltd.: Singapore; 2009; Vol. 1, pp 61–86.

10. Ramteke, M.; Gupta, S. K. Multi-Objective Genetic Algorithm and Simulated Annealing with the Jumping Gene Adaptations. In *Multi-Objective Optimization: Techniques and Applications in Chemical Engineering, Advances in Process Systems Engineering;* Rangaiah, G. P., Ed.; World Scientific Publishing Co. Pte. Ltd.: Singapore; 2009; Vol. 1, pp 91–129.

11. Miettinen, K.; Hakanen, J. Why use Interactive Multi-Objective Optimization in Chemical Process Design? In *Multi-Objective Optimization: Techniques and Applications in Chemical Engineering, Advances in Process Systems Engineering;* Rangaiah, G. P., Ed.; World Scientific Publishing Co. Pte. Ltd.: Singapore; 2009; Vol. 1, pp 153–187.

12. Garg, S. Array Informatics using Multi-Objective Genetic Algorithms: From Gene Expressions to Gene Network. In *Multi-Objective Optimization: Techniques and Applications in Chemical Engineering, Advances in Process Systems Engineering;* Rangaiah, G. P., Ed.; World Scientific Publishing Co. Pte. Ltd.: Singapore; 2009; Vol. 1, pp 363–396.

13. Andersson, J. Multi-Objective Optimization in Engineering Design: Applications to Fluid Power Systems, Dissertation, Linkoping Studies in Science and Technology, Linkoping University, Sweden, 2001.

14. Swan, R. W. Optimization of Water Treatment Works using Monte-Carlo Methods and Genetic Algorithms, Doctoral Thesis, University of Birmingham, 2015.

15. Dawood, A. S.; Li, Y. Modeling and Optimization of New Flocculant Dosage and pH for Flocculation: Removal of Pollutants from Wastewater. *Water* **2013**, *5*, 342–355.

16. Shau, H.-M.; Lin, B.-L.; Huang, W.-C. Genetic Algorithms for Design of Pipe Network Systems. *J. Mar. Sci. Technol.* **2005**, *13*(2), 116–124.

17. Sawaragi, Y.; Nakayama, H.; Tanino, T. *Theory of Multi-Objective Optimization, Mathematics in Science and Engineering;* Academic Press: USA; 1985.

CHAPTER 4

BEER, ALL A SCIENCE, ALCOHOL: ANALYSIS AND QUALITY TESTING AT MULTIPLE BREWERY STAGES

FRANCISCO TORRENS[1,*] and GLORIA CASTELLANO[2]

[1]*Institut Universitari de Ciència Molecular, Universitat de València, Edifici d'Instituts de Paterna, P. O. Box 22085, E-46071 València, Spain*

[2]*Departamento de Ciencias Experimentales y Matemáticas, Facultad de Veterinaria y Ciencias Experimentales, Universidad Católica de Valencia San Vicente Mártir, Guillem de Castro-94, E-46001 València, Spain*

*Corresponding author. E-mail: torrens@uv.es

ABSTRACT

Beer is good for health. In ancient Greece and throughout the middle ages, beer and ethanol were used for various medical applications. In recent decades, the appearance of powerful medicines eclipsed ethanol as a remedy. Recently, however, interest revived in ethanol's possible positive effects on health. Caffeine and alcohol are drug models because they are well-studied medicines. They are widely consumed psychoactive drug. Caffeine, alcohol, and theobromine may serve as lead compound for the development of novel drugs. The popularization of cafés during the 19th century, replacing beer bars, decreased the consumption of alcohol during working days, improving health and safety at work. The flow sheet of information with the application of analytical methods was informed.

4.1 INTRODUCTION

Beer and its history were reviewed by Belgian Brewers.[1]

In earlier publication, a comparison was made between beer bars cafes (19th century).[2-4] The aim of the present report is to analyze and discuss the commodities of the brewing process, the raw materials of the brewing process (water, malt, hops, and yeast), how to serve a beer, all a science, alcohol, analysis and quality testing at multiple brewery stages.

4.2 BREWING: THE COMMODITIES

1. The malt is milled and crushed to produce malt flour and stored in silos.[1]
2. Malt flour is poured into the mash tank with water. The mixture is heated, which dissolves soluble components of malt and converts starch into soluble sugars, which will ferment.
3. The sweet mixture is left to rest and filtered in order to get rid of the spent grains. The remaining liquid is called "wort," a clear solution of fermentable sugars. The desired quantity of hops is added to give the beer its specific aroma. The wort is boiled. After boiling, the hops are removed by sedimentation/centrifuging and the brew is cooled to the desired fermentation temperature.
4. The wort flows over cooling pipes or through a plate cooler before being aerated. Yeast is added to the wort in order to start the fermentation. Yeast converts the sugars into ethanol and CO_2, and the wort into beer. The wort is left to ferment for 2 weeks depending on the type of yeast.
5. Maturation: flavor, bouquet, and character are refined and ethanol content is determined.
6. The beer stays turbid because of the remaining yeast cells and other suspended particles. The last filtration will make the beer crystal clear. However, special beers remain unfiltered (cloudy).
7. The beer is ready for consumption but must be bottled, canned, or barreled.

4.3 BREWING PROCESS: RAW MATERIALS

Water: Water is an important element in brewing. Its mineral composition influences beer's color and taste.[1] The water used as a raw material for the beer is called "brewing water." Besides water, steam is used for the heating and cleaning of the installations.

Malt: Malt is barley that was dried and germinated. Barley grains germinate with water, warmth, and air. After the grain is moistened, it begins to germinate. The germs produce enzymes that convert the starch in the grain into sugars during brewing. As soon as the enzymes appear, germination slows down. Then, germination is stopped by kilning the grain. The higher the kilning temperature, the darker the malt and beer. Besides enzymes and starches, malt contains proteins, which will give the beer all its tasting extent and beautiful froth.

Hops: Hops give beer its typical fine bitter taste and flavor. Hop varieties can be divided into two classes: bittering and aroma hops, which contain a high level of aromatic volatile oils.

Yeast: Yeast is a microorganism that converts the sugars into ethanol and CO_2. Every beer has its yeast culture. The yeast is propagated from one "single mother cell." In order to guarantee a constant taste and flavors, the yeast is always propagated from the master culture. Depending on the temperature and the yeast, three types of fermentation exist: top, bottom, and spontaneous.

4.4 HOW TO SERVE A BEER?

One has different beers (*cf.* Table 4.1); that is, for everyone and all occasions.[1]

Every beer presents its proper glass and is served according to the rules of the art.

1. Bottled beer should be stored in a dark and dry place.
2. Cool the beer: place the bottles in the refrigerator at least 24 h before serving.
3. Temperature: "thirst-quenching beers": 3°C; "gourmet beer": 6–8°C.
4. Clean glasses with cold water in which a good detergent is dissolved and rinse thoroughly with water. Glasses for gourmet beers should be dried.

5. Pouring: "thirst-quenching beer": pour the beer all in one go. Tip the glass slightly to one side and raise it gradually to an upright position. Let the froth flow over the sides and then skim off the surface bubbles of the froth with a clean knife. Rinse the outside of the glass. "Gourmet beer": serve the beer slowly as to create a rich foamy head. Leave some beer in the bottle so the glass can be topped up afterward. For beers that are bottle conditioned, leave the yeast deposit in the bottom of the bottle and present the bottle with the glass.

TABLE 4.1 Beer Types Brewed in Belgium and Fermentation Methods.

Beer types brewed in Belgium	Fermentation methods
International	
Pils	Bottom fermentation
Premium Pils	Bottom fermentation
Light	Bottom fermentation
Low alcohol and nonalcoholic	Bottom/top fermentation
English style	Top fermentation
Belgian specialities	
Amber	Top fermentation
White	Top fermentation
Abbey	Top fermentation
Trappist	Top fermentation
Gueuze-Lambic	Spontaneous fermentation
Fruit	Spontaneous and/or top fermentation
Strong blond	Top fermentation
Regional	Top fermentation
Red brown (acid)	Mixed fermentation
Table	Bottom/top fermentation

Source: Adapted from ref 1.

Cheers!

Successful combinations of beer and dishes exist (*cf.* Table 4.2).

TABLE 4.2 Successful Combinations of Beer and Dishes.

Beer type	Appetizer	Seafood	Cooked fish	Boiled fish	Assorted delicacy	White meat (poultry)	Red meat	Game	Light cheese	Strong cheese	Desserts
Alcohol-free	X	X	X	X	X	X	X		X		
Pils	X	X	X	X	X	X	X		X		
Premium pils	X	X	X	X	X	X	X		X		
White beer	X	X	X	X	X	X	X		X		
Gueuze	X	X	X	X	X	X				X	
Sour beer	X	X	X	X	X	X				X	X
Amber	X	X	X	X	X	X				X	X
Abbey + Trappist	X[1]			X	X		X	X		X	X
Strong blond				X	X		X	X		X	X
Scotch					X		X	X		X	X
Fruit beer	X										X

[1] Blond abbey beer.

Source: Adapted from ref 1.

4.5   BEER AND SCIENCE: BEER, ALL A SCIENCE

Collective Innocampus organized a Round Table on Beer and Science with tasting and gastronomic combination.[5] Pedrós proposed Qs/As on physics of beer.

Q1. Can all knowledge of physics be applied to beer?

A1. For instance,

$$\text{responsible density} = \frac{\text{mass of bear}}{\text{responsible volume}} \qquad (4.1)$$

However, "responsible volume" depends from one to another person and on number of drunken beers.

Q2. To use physics to explain beer?

Q3. Is the skunk smell of beer due to green or brown bottles a scientific explanation?

A3. Yes, ultraviolet radiation causes the degradation of isohumulones.

Q4. What is mean radius r of bubbles issuing from a small area (spur, dust, defect) of glass surface?

A4. The mean bubble radius r is:

$$r = \sqrt[3]{\frac{3R\sigma}{2\rho g}} \qquad (4.2)$$

where R is nucleation site radius; σ, beer surface tension; ρ, beer density; g, gravity acceleration.

Q5. How long have humans been drinking beer?

A5. Archeology says 10 years, but the mutation in the enzyme degrading ethanol is much older!

Q6. How long have humans been manufacturing beer?

Q7. Will people drink beer in the future?

Q8. Can one drink beer in the weightlessness?

A8. Yes, through a straw, but belching causes one to vomit!

Hipólito raised questions on the effect of alcohol in organism and responsible consumption.[6]

Q9. What is the effect of drinking compulsively alcohol in the organism in the long term (*cf.* Fig. 4.1)?

Q10. What is it equivalent to every glass in units of alcohol (1 unit = 10 ml = 8 g pure alcohol) for?

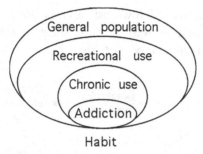

Habit

FIGURE 4.1   Decay of population involved in habit (recreational use: 10–15%).

4.6   ALCOHOL: ANALYSIS AND QUALITY TESTING AT MULTIPLE BREWERY STAGES

Ktori raised the following hypothesis, questions and answers on alcohol and analysis.[7]

H1. Dedicated brewery management systems do not have the capacity to handle the data from laboratory-based quality/testing workflows.

Q1. What does not brewery management software provide?

A1. It does not provide any scope for managing quality control testing and resulting data/reporting.

Q2. Why, if so much testing must be performed as a matter of course, do not brewery management packages include, as standard, capacity to manage quality-related testing/data?

A2. Managing laboratory processes is a harder technical nut to crack than customer administration functions.

Q3. How best to manage, streamline and provide transparency for industry laboratory functions?

Q4. How does our software work?

Q5. What are the trends?

Q6. Why not to implement corrective measures?

Q7. Where is every beer in the brewery?

Q8. How much of every beer is it in the brewery?

Q9. What does every tank contain every day?

De la Guardia reviewed low-cost devices for improving screening analytical methods and showed the flow sheet of information with the application of analytical methods (*cf.* Fig 4.2).[8]

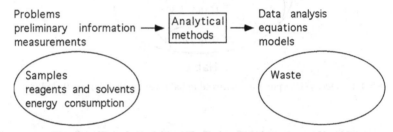

FIGURE 4.2   Flow sheet of information in analytical chemistry.

4.7   DISCUSSION

Beer is good for health. In ancient Greece and throughout the Middle Ages, beer and ethanol were used for various medical applications (e.g., treating fevers, wounds, and sleep disorders). In recent decades, the appearance of powerful medicines eclipsed ethanol as a remedy. Recently, however, interest revived in ethanol's possible positive effects on health.

The popularization of cafes during the 19th century, replacing beer bars, decreased the consumption of alcohol during working days, improving health and safety at work.

De la Guardia showed the flow sheet of information with the application of analytical methods.

4.8   FINAL REMARKS

From the present results and discussion, the following final remark can be drawn.

Beer is good for health. In ancient Greece and throughout the Middle Ages, beer and ethanol were used for various medical applications. In recent decades, the appearance of powerful medicines eclipsed ethanol as a remedy. Recently, however, interest revived in ethanol's possible positive effects on health. Caffeine (from coffee, tea, etc.) and alcohol are drug models because they are well-studied medicines. They are widely consumed the psychoactive drug. They and theobromine (from cocoa) may serve as lead compound for the development of novel drugs.

ACKNOWLEDGMENT

Francisco Torrens belongs to the Institut Universitari de Ciència Molecular, Universitat de València. Gloria Castellano belongs to the Departamento de Ciencias Experimentales y Matemáticas, Facultad de Veterinaria y Ciencias Experimentales, Universidad Católica de Valencia San Vicente Mártir. The authors acknowledge the support from Generalitat Valenciana (Project No. PROMETEO/2016/094) and Universidad Católica de Valencia San Vicente Mártir (Projects No. 2017).

KEYWORDS

- alcohol
- analytical method
- brewing process
- commodities
- ethanol
- raw material
- serving beer

REFERENCES

1. Brewers, B. *Brewers' House;* Belgian Brewers: Brussels, 2017.
2. Torrens, F.; Castellano, G. QSPR Prediction of Retention Times of Methylxanthines and Cotinine by Bioplastic Evolution. *Int. J. Quant. Struct. Prop. Relat.* **2018**, *3*, 74–87.

3. Torrens, F.; Castellano, G. Molecular Classification of Caffeine, its Metabolites and Nicotine Metabolite. In *Frontiers in Computational Chemistry;* Ul-Haq, Z., Madura, J. D., Eds.; Bentham: Hilversum, Holland, (in press).

4. Torrens, F.; Castellano, G. QSPR Prediction of Retention Times of Chlorogenic Acids in Coffee by Bioplastic Evolution. In *Quantitative Structure-Activity Relationship;* Kandemirli, F., Eds.; InTechOpen: Vienna, (in press).

5. Innocampus, *Book of Abstracts, Beer and Science, València, Spain, March 22, 2017;* Universitat de València: València, Spain, 2017; MR–M1.

6. Preedy, V. (Ed.). *Beer in Health and Disease Prevention;* Elsevier: Amsterdam, 2008.

7. Ktori, S. Alcohol and Analysis. *Sci. Comput. World* **2017**, *2017*(153), 18–20.

8. De la Guardia, M. *Book of Abstracts, X International Workshop on Sensors and Molecular Recognition, València, Spain, July 7–8, 2016;* Universitat de València: València, Spain, 2016; CP–C2.

CHAPTER 5

OZONATION OF INDUSTRIAL WASTEWATER AND THE FUTURE OF INDUSTRIAL POLLUTION CONTROL: A CRITICAL REVIEW AND A VISION FOR THE FUTURE

SUKANCHAN PALIT[1,2]

[1]Department of Chemical Engineering, University of Petroleum and Energy Studies, Bidholi via Premnagar, Dehradun 248007, India

[2]43, Judges Bagan, Haridevpur, Kolkata 700082, India

E-mail: sukanchan68@gmail.com; sukanchan92@gmail.com, Mob.: 0091-8958728093

ABSTRACT

The world of environmental engineering is witnessing one paradigmatic shift in science over another. The frequent environmental disasters, the stringent environmental regulations and the grave concerns of environmental sustainability have urged the scientific domain to move toward newer technologies and newer innovations. Technology and engineering science today stands in the midst of deep vision and introspection. Technological vision and scientific objectives of ozone oxidation of industrial wastewater need to be reenvisioned and readdressed at each step of scientific endeavor. Hazardous organic wastes from industrial, military, and commercial operations represent one of the foremost challenges to environmental engineers. This chapter broadly discusses the immense potential of advanced oxidation techniques primarily ozonation in the treatment of industrial wastewater. Advanced oxidation processes (AOPs) are viable alternatives to the incineration of wastes, which has many disadvantages. This chapter provides a detailed overview of theoretical basis, efficiency, economics, design, and modeling of different AOPs (combinations of ozone and hydrogen peroxide with ultraviolet (UV) radiation and catalysts). The author of this chapter pointedly focuses on the immense success and tremendous viability of scientific endeavor in the field of ozone oxidation of industrial wastewater. The scientific vision and the deep scientific comprehension of ozone reactions with inorganic and organic compounds are delineated in details in this chapter. Ozone is a powerful oxidizing agent. Second only to fluorine is the oxidizing power of the ozone and has many uses including but not limited to water purification, bleaching of materials such as paper, synthetic fibers, Teflon, waxes, flour, and other products, treatment of wastes in industry, deodorization, and sterilization. Ozone is a safe alternative to chlorine products which performs the same functions without the undesirable side effects; it is not harmful to the environment since it is made from oxygen and decomposes to oxygen. Scientific introspection, the deep scientific insight, and the futuristic challenges will go a long way in the true emancipation of environmental engineering techniques.

5.1 INTRODUCTION

Environmental engineering science is witnessing drastic and dramatic challenges today. Frequent environmental disasters, the breach in industrial pollution control, and the loss of ecological biodiversity have plunged human civilization and human scientific endeavor toward newer innovations and newer technologies. The challenge and the vision of environmental engineering are absolutely inspiring as science crosses wide and visionary frontiers. This chapter treads a weary path toward scientific destiny and scientific vision in the field of ozone oxidation (ozonation) and other integrated AOPs. Technology and science of environmental engineering today stands in the midst of deep scientific introspection and scientific comprehension. Novel separation techniques and nontraditional environmental engineering techniques are gaining immense scientific heights. Science is a huge colossus with a definite vision and will of its own. The author, in this chapter, pointedly focuses on the success of ozonation techniques in solving industrial water pollution control problems.

Environmental sustainability and industrial wastewater treatment issues are the opposite sides of the visionary coin. Chemical process engineering and environmental engineering science are the forerunners and torchbearers toward a visionary future of science and technology. In the similar fashion, novel separation processes such as membrane science are changing the scientific landscape of scientific research pursuit. Environmental regulations and immense global environmental restrictions have plunged the scientific domain toward newer innovations and new scientific instincts. This chapter widely observes the success of novel separation techniques and nontraditional environmental engineering processes in the furtherance of science and engineering.[1,2]

5.2 AIM AND OBJECTIVE OF THIS STUDY

The vision of this short study goes beyond scientific understanding and is replete with scientific ingenuity and scientific profundity. Today, environmental engineering science stands tall in the midst of immense scientific rejuvenation and immense scientific challenges. The environmental catastrophes and the challenges of industrial water pollution control will go a long and visionary way in the true emancipation of environmental engineering science today. Hazardous organic wastes from industrial, military, and commercial operations represent drastic challenges to the environmental engineers and environmental scientists. This chapter rigorously focuses on the different AOPs especially ozone oxidation or ozonation. The central objective of this chapter is vast and versatile. Environmental engineering science and chemical process engineering are witnessing the challenge of our times. Industrial wastewater treatment, water pollution control, and water purification will inevitably go a long way in the true emancipation of environmental engineering science and environmental sustainability in years to come. Environmental engineering is in a state of immense disaster. The integration of advanced oxidation technologies (AOTs) and other traditional wastewater treatment processes have been proven to be more effective for treating polluted water—drinking water and industrial wastewater. Technology is emboldened and scientific objectives are realized as science moves toward a newer scientific regeneration.[1,2]

5.3   NEED AND THE RATIONALE OF THIS STUDY

The need and the rationale of this study are immense and far-reaching. Scientific profundity, scientific vision, and immense scientific forbearance are the torchbearers toward a greater emancipation of environmental science. The science of environmental engineering is witnessing dramatic changes as sustainability issues gain new heights. Holistic sustainability development and sound traditional and nontraditional environmental engineering techniques are the two opposite sides of the visionary scientific coin. Breach of industrial pollution control and the loss of environmental and ecological biodiversity are the forerunners toward a greater realization and greater application of environmental techniques. AOPs and the relevant area of ozonation or ozone oxidation are changing the face of industrial wastewater treatment and the immensely relevant field of drinking water treatment. Water purification is a pillar and cornerstone of environmental engineering endeavor. This urge to excel in science and engineering has plunged scientific domain in the quest toward zero-discharge norms. Scientific notion and scientific understanding are witnessing difficult challenges today due to the proliferation of nuclear science and least scientific regard toward environmental engineering norms. In such a crucial juncture of scientific history and time, scientific vision and technological objectives are gaining immense scientific heights. The author repeatedly stresses, in this chapter, on the scientific doctrines of AOPs, primarily ozonation, and this treatise goes beyond scientific imagination and scientific judgment. The success of environmental engineering science is reviewed with scientific skill and scientific candor. Technology and engineering science are veritably emboldened as the scientific domain ushers in a new era. Industrial wastewater treatment is revitalizing the science of environmental engineering. The need and the rationale of this study are immense and far-reaching. The question of environmental biodiversity and environmental balance has urged the scientific domain to move toward newer innovations and newer technological forays. In this chapter, the author pointedly focuses on the deep scientific success, the scientific forbearance, and the intellectual prowess behind ozone oxidation applications in industrial wastewater treatment.[1,2]

5.4 OZONE OXIDATION

Technology of ozone oxidation is vast, versatile, and far-reaching. Industrial wastewater treatment and water purification issues are enigmatic to the furtherance of science and technology. The vision and challenge of science are changing the face of scientific endeavor in the field of ozone oxidation. Industrial water pollution control today stands in the midst of immense scientific vision and deep scientific introspection. Technology needs to be reenvisioned and reenshrined as human civilization moves toward a newer visionary eon. The question of zero-discharge norm is perplexing the entire scientific domain. This chapter repeatedly stresses on the futuristic vision and the future trends in research in the field of AOPs, primarily ozone oxidation.

The oxidation of organic and inorganic compounds during ozonation can occur through ozone or OH radicals or a combination thereof. The oxidation pathway is determined by the ratio of ozone and OH radical concentrations and the corresponding kinetics. Ozone is an electrophile with a high selectivity.[1,2] The reactions of ozone with drinking-water-relevant inorganic compounds are typically fast and occur by an oxygen atom transfer reaction. Organic micropollutants are oxidized with ozone selectively. Ozone reacts mainly with double bonds, activated aromatic systems, and non-protonated amines. In general, electron-donating groups enhance the oxidation by ozone, whereas electron-withdrawing groups reduce the reaction rates.[1,2]

5.5 ADVANCED OXIDATION PROCESSES (AOPS)

Advances in chemical water and wastewater treatment have led to the development of methods termed AOPs or AOTs. AOPs can be broadly defined as aqueous phase oxidation methods based on the intermediacy of highly reactive species such as hydroxyl radicals in the mechanisms leading to the destruction of the targeted pollutants. Over the past 30 years, R&D concerning AOPs has been immensely particular for two reasons, namely the wide diversity of technologies involved and the areas of potential applications. Key AOPs include heterogeneous and homogeneous photocatalysis based on near-UV or solar visible irradiation, electrolysis, ozonation, Fenton's reagent, ultrasound and wet air oxidation (WAO), while less conventional but evolving processes include ionizing radiation, microwaves, pulsed plasma, and ferrate reagent.[1,2]

Technology and science of AOPs are today surpassing vast and versatile visionary frontiers. Environmental engineering science today stands in the midst of immense scientific comprehension and deep technological insight. The challenge and vision of AOPs and ozonation are immense and far-reaching. The scientific urge to excel, the definitive vision, and the wide scientific rigor will all lead to a long way in the true emancipation of environmental engineering and environmental science. Today, science is a visionary domain of human endeavor. Advances in human civilization, the immense academic rigor, and the global water issues are urging the scientific world to gear toward newer technologies and newer innovations.[1,2]

5.6 RECENT SCIENTIFIC ENDEAVOR IN THE FIELD OF OZONE OXIDATION

Ozone is a strong oxidant and an excellent disinfectant. Ozone is widely used in water treatment plants throughout the world to address disinfection, disinfection by-products, taste and odor, color, micro-coagulation, and the other water treatment needs and necessities. Ozone can effectively eliminate bacteria and inactive viruses more rapidly than any other disinfectant chemicals.[1,2] Technology of ozone oxidation is highly advanced today. Ozone reacts to oxidize a number of inorganic compounds including iron, manganese, sulfides, nitrite, arsenic, bromide ion, and iodide ion.[1,2]

Ozone oxidation today is on the path of immense scientific vision and deep scientific understanding. The state of the environment is at a devastating state. Nontraditional environmental engineering techniques are the utmost and the immediate need of the hour. Technological objectives and scientific candor are changing the face of technological and scientific innovation. The prime need of the hour is drinking water treatment, water purification, and industrial wastewater treatment. The wide visionary domain of chemical process engineering and environmental engineering is veritably in the path of new rejuvenation.[1,2]

Augustina et al.[3] reviewed synergistic effect of photocatalysis and ozonation on wastewater treatment. For the treatment of wastewater that contains recalcitrant organic compounds, such as organo-halogens, organic pesticides, surfactants, and coloring matter, wastewater engineers are now required to develop AOP.[3] A promising and effective way to perform mineralization of this type of substance is the widespread application of AOP.[3] Technology needs to be reenvisioned as human civilization and human scientific endeavor move toward a newer era of scientific regeneration and scientific vision. The challenge of science is awe-inspiring as engineering science as well as environmental engineering moves from one scientific genre toward another. Science, engineering, and technology are moving in visionary directions. Photocatalysis and ozonation come within the boundary of integrated AOPs.[3] Thus, it is of immense importance and vital in scientific understanding.

Rosal et al. (2010)[4] rigorously delineated in a visionary treatise the occurrence of emerging pollutants in urban wastewater and their removal through biological treatment followed by ozonation.[4] Technological advancements and scientific grandeur are immensely enhanced as ozonation opens up wide windows of scientific innovation and instinct. The work widely presented a systematic survey of over 70 individual pollutants in a sewage treatment plant receiving urban wastewater.[4] The challenge of technology in ozonation science and integrated environmental engineering techniques (traditional or nontraditional) is gaining immense heights as scientific endeavor gears forward. The compounds treated include mainly pharmaceuticals and personal care products (PPCPs) as well as some metabolites. The quantification was performed by liquid chromatography–mass spectrometry and gas chromatography coupled with mass spectrometry.[4] A kinetic model was analyzed to determine the second-order kinetic constants for the ozonation of bezafibrate, cotinine, diuron, and metronidazole.[4] The experimental results showed that the hydroxyl radical reaction was the major pathway for the oxidative transformation of these compounds.[4]

Palit[5] discussed lucidly ozone oxidation of dye in a bubble column reactor at different pH and different oxidation–reduction potentials. The dye-effluent waste cannot be degraded with the help of primary and secondary treatments, thus the need of tertiary treatment such as AOPs or ozonation.[5] Scientific research pursuit in operation of bubble column reactor is highly advanced and ground-breaking. Intellectual prowess, scientific genesis, and deep scientific determination are the veritable needs of research pursuit in bubble column reactor. This chapter focuses on the understanding of the dependence of order of reaction on pH of the solution and oxidation–reduction potential.[5] This technology is new and highly advanced. The research on ozonation of dye was done in a bubble column reactor fixed bed and without media. Scientific prowess and deep scientific introspection are the imperatives of research pursuit today. This chapter redefines the scientific potential, the scientific girth, and the vision to move forward in bubble column reactor dynamics.[5] This research work also gives an understanding and an innovative concept that dye degradation by ozone is highly dependent on the acidity and alkalinity of the solution. Science and technology are surpassing vast and versatile frontiers. Technological girth and prowess need to be reenvisioned and reorganized as regards to fluid dynamics of bubble column reactor.[5] This chapter vastly reorganizes and redefines the human scientific endeavor in environmental engineering.[5]

Guo et al.[6] discussed with deep foresight the application and reaction mechanism of catalytic ozonation in water treatment. PPCPs are considered as the emerging environmental problem in the recent years. The elimination of PPCPs during water treatment is deeply investigated in the laboratory through different AOPs which are the technologies based on the application of hydroxyl and other radicals to oxidize inorganic and organic pollutants.[6] Technology and engineering science are today in a state of immense vision and fortitude. Catalytic ozonation is proved as an effective technology for the removal of organics from wastewater. The research paper vastly elucidated a short review about the introduction of the ozonation catalysts and the intricate reaction mechanism of the catalytic ozonation. The main vision is to provide a new and effective technique for the removal of PPCPs in the aqueous solution.[6] Human scientific regeneration and deep scientific determination in environmental engineering science are in a state of revival today.[6] Technological validation of AOP applications needs to be reenvisioned and restructured as science and engineering march forward toward a newer scientific eon. This chapter opens up new scientific frontiers and far-reaching visionary boundaries in environmental engineering in decades to come.[6]

Eriksson[7] elucidated in her licentiate thesis ozone chemistry in aqueous solution along with ozone decomposition and stabilization. Ozone is used in many applications in the industry as an oxidizing agent, for example, for bleaching and disinfection. The decomposition of ozone in aqueous solutions is highly complex and is affected by many properties such as pH, temperature, and substances present in the water.[7] The author in the treatise discussed in details general properties of ozone in aqueous solutions, experimental methods, kinetic experiments, ozonization, nuclear magnetic resonance spectroscopic analysis, decomposition rate in acidic aqueous solutions, and micellar enclosure of ozone.[7] Ozone stabilization is a scientific imperative in environmental engineering science. Technological validation, scientific motivation, and the deep scientific rigor are the forerunners toward a greater scientific emancipation in ozone chemistry and the research endeavor.[7] The aim of this project was to understand the fundamental ozone chemistry and to investigate stabilization of ozone and aqueous solutions. The main work is focused on the possibilities to stabilize ozone by micellar enclosure including kinetic studies.[7]

Zhu et al.[8] discussed with immense foresight the catalytic ozone oxidation with nano-TiO_2 modified membrane for the treatment of municipal wastewater. This area of research pursuit is unexplored yet far-reaching. Ozone aeration can produce a large number of high oxidizing free radicals to degrade organic matters when ozone comes in contact with nano-TiO_2 modified membrane which is called catalytic ozone oxidation processes.[8] This approach is a new scientific frontier of municipal wastewater treatment, which has excellent reactive activity and degradation of organic compounds without the need of the catalyst recycling. Research frontiers today in the combination of membrane science and catalytic ozonation processes are gaining immense heights with the progress of scientific and academic rigor.[8] The main mission of the research work was to remove organic matters by ozone aeration pretreatment and nano-TiO_2 modified membrane. The challenge and the vision of science in membrane separation phenomenon are wide and far-reaching. This chapter gives wide glimpses in the scientific endeavor in integrated membrane science technique and ozonation technique.[8]

Broseus et al.[9] pointedly focuses on the ozone oxidation of pharmaceuticals, endocrine disrupters, and pesticides during drinking water treatment. The challenges in drinking water treatment are vast and versatile.[9] Human scientific endeavor are today on the path toward new scientific regeneration and deep vision. PPCPs, endocrine-disrupting compounds (EDCs), and pesticides are groups of micropollutants which are routinely detected in surface waters. This is an area of immense concern. A significant fraction of the PPCP and EDCs released into the aquatic environment is attributed to their complete elimination through conventional wastewater treatment, thus the need for this extensive study. The most representative pharmaceutical compounds detected in urban wastewaters are anti-inflammatory drugs, anticonvulsants, antibiotics, and lipid regulators. This visionary scientific endeavor is slowly opening up new windows of scientific innovation in decades to come.[9] Scientific vision, scientific forbearance, and fortitude are the necessities of research pursuit today. With vast scientific imagination, the paper rigorously pointed out toward the efficacy of ozone oxidation of recalcitrant chemical compounds.[9]

Liu et al.[10] deeply comprehended with immense foresight the removal of trace antibiotics from wastewater in a systematic study of nanofiltration (NF) combined with ozone-based AOPs.[10] The work deeply investigated the removal of trace antibiotics from wastewater treatment plant (WWTP) effluent through NF and the disposal of the NF concentrate by AOPs.[10] Four antibiotics, namely norfloxacin, ofloxacin, roxithromycin, and azithromycin, which had high detection frequencies in WWTPs in Dalian (China) were selected as veritable micropollutants.[10]

Ozonation or ozone oxidation is one of the primary branches of AOPs today. Industrial wastewater treatment and drinking water treatment stands in the midst of deep vision, introspection, and scientific revival. The entire scientific domain is focused on the success of energy and environmental sustainability. The domain of environmental sustainability has an unsevered umbilical cord with environmental engineering science. The authors in the well-researched treatise pointedly focused on the vast scientific panorama and the scientific landscape in the area of AOPs with the sole aim of furtherance of science and engineering.

5.7   VISIONARY SCIENTIFIC RESEARCH PURSUIT IN THE FIELD OF AOPS

AOPs today stands in the midst of deep scientific comprehension and wide visionary introspection. The challenge and the deep vision of integrated AOPs are ushering in a new eon in the field of environmental engineering scenario. Industrial wastewater treatment scenario and drinking water treatment are facing immense challenges. Global environmental sustainability is on the path toward newer scientific regeneration. Scientific vision and scientific candor in the field of nontraditional environmental engineering techniques are witnessing dramatic challenges. In this chapter, the author pointedly focuses on the success of research and development initiatives in AOPs and integrated AOPs in the furtherance of science and ushering in a new path scientific glory and scientific profundity. Scientific forbearance and scientific vision are of utmost need in the progress of scientific and academic rigor in the field of AOPs. Drinking water treatment and industrial wastewater treatment are the need of the hour as science and engineering enter into a newer age. Environmental sustainability and its scientific vision and deep scientific discernment are the necessities of present-day scientific research pursuit.

Munter[11] discussed with deep and cogent insight current status and prospects in AOPs. The paper presented an overview of theoretical basis, efficiency, economics, laboratory and pilot plant testing, and design and modeling of different AOPs (combinations of ozone and hydrogen peroxide with UV radiation and catalysis).[11] Technological vision, scientific motivation, and the challenge of scientific validation will all lead to a long and visionary way toward the true emancipation of environmental sustainability.[11] Hazardous organic wastes from industrial, military, and commercial applications represent one of the greatest challenges to environmental engineers and scientists.[11] AOPs are the veritable alternatives to the incineration of industrial wastes, which has many disadvantages. The scientific success, the scientific fortitude, and the vast scientific girth are the forerunners toward a newer eon of science and engineering of nontraditional environmental engineering techniques.[11] Conventional incineration is deeply thought to be a feasible alternative to landfill, but as presently practiced, incineration can bring about serious problems due to releasing of toxic compounds such as polychlorinated dibenzodioxins and polychlorinated dibenzofurans into the incinerator of gas emissions and fly ash.[11] Today, scientific cognizance and scientific forays stand in the midst of vision and deep pragmatism. This challenge of science needs to be reenvisioned and reorganized with the passage of scientific history and time. The author in the well-researched treatise discussed with deep and widening foresight the scientific success of AOPs applications.[11] The AOPs have proceeded along one of the two routes. These routes are: oxidation with oxygen in temperature ranges intermediate between ambient conditions and those found in incinerators, WAO processes in the range of 1–20 MPa, and 200–300°C; the use of high-energy oxidants such as ozone and hydrogen peroxide and/or photons that are able to generate highly reactive intermediates—OH radicals.[11]

The AOPs according to a visionary research work can be defined as "near-ambient temperature and pressure water treatment processes which involve the generation of hydroxyl radicals in sufficient quantity for effective water purification." Technological vision, scientific ingenuity, and scientific comprehension are in a state of immense revamping and deep scientific discernment. Human civilization and human scientific endeavor are highly challenged and veritably replete with vision and scientific adjudication. Water treatment and industrial wastewater treatment stand in the midst of immense scientific catastrophe and vast scientific adjudication. Munter[11] successfully proclaims the vast importance of AOPs with the sole vision of advancement of science and engineering. In AOP, the hydroxyl radical (OH) is a powerful nonselective chemical oxidant. This technology is far-reaching and deeply reenvisioned. This OH radical acts very rapidly with most organic compounds.

Sharma et al.[12] redefine the domain of AOPs and their vast applications and scientific potential. The treatise is a general review on AOPs for wastewater treatment. Scientific vision, scientific profundity, and scientific forbearance are the vast forerunners toward a newer emancipation in nonconventional environmental engineering techniques. Environmental engineering science and chemical process engineering are two opposite sides of the visionary coin.[12] AOPs constitute a promising and far-reaching technology for the treatment of wastewaters containing non-easily removable organic compounds. All AOPs are designed to produce hydroxyl radicals.[12] It is the hydroxyl radicals that act with high efficiency to destroy organic compounds. The challenge, the vision, and the scientific prowess in the field of environmental engineering science are the torchbearers toward a newer visionary eon in nontraditional tools in industrial wastewater treatment. AOP combines ozone (O_3), UV, hydrogen peroxide, and/or catalyst to offer a powerful water treatment solution for the effective reduction and removal of residual organic compounds as measured by chemical oxygen demand, biochemical oxygen demand, and total organic carbon.[12] Scientific vision, scientific cognizance, and scientific ingenuity are the technological forerunners of advancement of AOPs in industrial wastewater treatment. In the well-researched treatise, fundamentals and applications of typical methods such as Fenton, electro-Fenton, photo-Fenton, ozonation, and UV radiation are discussed with lucid details. Human scientific rigor today is in a state of immense scientific regeneration.[12] The main goals of academic, research, and human society through the development and implementation of environmental applications of AOPs will be as follows:

1. New concepts, processes, and technologies in wastewater treatment with potential benefits for the stable quality of effluents, energy and operational cost savings, and the effective protection of the environment.[12]
2. New sets of advanced standards for wastewater treatment.[12]
3. New techniques for the redefinition of wastewater treatment needs and proper technical framework.
4. Enhancing the water industry competitiveness.[12]

Sharma et al.[12] deeply comprehended the scientific success of Fenton process, electro-Fenton process, and advantages and disadvantages of AOPs. AOPs seem to be environment-friendly processes for decolorization of real dyeing wastewater. AOPs represent a powerful mean for the abatement of refractory and toxic pollutants in wastewater.[12] Scientific forbearance and deep scientific discernment are gaining immense heights as challenges and vision of AOPs increase. Human scientific genre in AOP and other nontraditional techniques are changing the face of scientific research pursuit today. The authors reviewed the entire gamut of AOPs or integrated AOPs.[12]

Jelonek et al.[13] discussed with deep and cogent insight the use of AOPs for the treatment of landfill leachate. Technological motivation and scientific validation are regaining themselves today as science and engineering march forward. The formation of leachate poses a problem closely related to the use of landfill sites.[13] Landfill leachate is a wastewater, which as a result of permeation elutes mineral and organic compounds from a bed. The article showed that there is a variety of pollutants in municipal landfill leachate. It also presented the dependency existing between the age of a landfill site and the concentration of pollutants in leachate. This is an innovation and promising area of scientific endeavor. The vision and the challenge of scientific research pursuit are slowly evolving as science and technology move forward.[13] Various methods to purify leachate and the physical, chemical, physicochemical and biochemical processes were deeply compared. Scientific endeavor is slowly moving toward newer knowledge dimensions and in the area of AOPs, surpassing vast scientific frontiers.[13] The treatment of landfill leachate is one of the major scientific imperatives of AOPs today. Human civilization and human scientific endeavor in environmental engineering science stand in the midst of deep scientific comprehension and ingenuity. The authors in the treatise deeply elucidated the vast scientific success, the deep scientific potential, and the vast scientific profundity on the path toward scientific emancipation of environmental engineering. AOPs are divided into chemical and photo-chemical oxidation. The most commonly used methods include oxidation with ozone and hydrogen peroxide and Fenton's reagent oxidation. Scientific endeavor and scientific profundity are the forerunners toward a newer visionary eon in environmental engineering science. This chapter rigorously pointed out toward the rigors of science and engineering of the environment on the human planet.[13]

Gilmour[14] discussed with deep and cogent insight the application perspectives in water treatment using AOPs. This is a watershed text in environmental engineering science. AOPs using hydroxyl radicals and other oxidative radical species are being studied immensely with scientific vision and scientific determination in the present-day human scientific domain.[14] Technology and engineering science are today in a state of immense scientific regeneration. Large-scale applications of AOPs in industrial wastewater treatment and drinking water treatment are highly limited due to cost and inadequate information of water quality. This comprehensive study focused on upstream processing and downstream posttreatment analysis of selective AOPs.[14] Human scientific rigor and human scientific research pursuit are today in a state of deep comprehension.[14] The challenge and the vision need to be restructured and reorganized with every step of human scientific history. In the first stage of the research, the performance of a proprietary catalyst was compared with the industry standard for the use in a pilot-scale immobilized photocatalytic reactor. Scientific regeneration and scientific rejuvenation are today in a state of immense vision and forbearance. In the second stage of the study, two bioassays were used to evaluate and compare the toxicity of bisphenol A and its degradation intermediates formed in the three AOPs, namely UV/hydrogen peroxide, ozonation, and photocatalysis.[14]

5.8   SCIENTIFIC VISION AND SCIENTIFIC DOCTRINE OF INDUSTRIAL WASTEWATER TREATMENT AND AOPs

AOPs and ozone oxidation are culminating in a newer visionary era of environmental engineering and environmental sustainability. Technology and science are today challenged as human scientific endeavor moves from one paradigm toward another. Nanoscience and nanotechnology are the two opposite sides of the visionary coin. Chemical process engineering in the similar vein is connected by an unsevered umbilical cord with environmental engineering science. The scientific doctrine of chemical process engineering is today linked with AOPs. Environmental awareness and subsequent environmental regulations are urging the world of science and engineering to gear forward for new technological innovations. Science and technology are today huge colossus with a vast vision and prowess of its own. Industrial wastewater treatment is a vexing issue troubling the scientific domain and the human civilization. The definite and futuristic vision of industrial wastewater treatment technologies involves the development of innovative technologies and gear forward toward zero-discharge norms. In such a crucial juxtaposition of scientific juxtaposition, innovative technologies such as AOPs need to be reenvisioned and reenvisaged. The challenges, the vision, and the targets of science and engineering will go a long and visionary way in the true realization of environmental sustainability and environmental engineering science. Human civilization today stands in the midst of vast scientific vision and fortitude. Environmental engineering and chemical process engineering are today veritably linked by an unsevered umbilical cord. In this chapter, the author repeatedly stresses on sustainable development, the vast arenas of industrial wastewater treatment, and global water issues.[15,16,17]

5.9 DRINKING WATER TREATMENT AND THE PROGRESS OF SCIENCE

Human civilization and human scientific endeavor today stand in the midst of deep vision and a greater scientific discernment. Provision of pure drinking water is a challenge to the vast global scientific scenario. Today, global water shortage and groundwater heavy metal contamination are the enigmatic issues facing human scientific endeavor. Environmental engineering techniques—traditional as well as nontraditional are opening up new vistas in scientific research pursuit. The intricate challenges in global water research and development initiatives need to be redefined and reshaped as science and engineering crosses one visionary boundary over another. Arsenic groundwater contamination in many countries of the world is seriously devastating the scientific landscape. Technology and engineering science are in the throes of immense scientific difficulties and barriers. Progress of science and technology and the new beginning of engineering science will lead to a long and visionary way in the true visionary realization of water science and technology. Technological diversification in water science is the need of the hour as the human scientific endeavor moves forward.[15,16,17]

5.10   VISION OF INDUSTRIAL WASTEWATER TREATMENT

Industrial wastewater treatment and drinking water purification technologies today stands in the midst of immense scientific comprehension and deep scientific cognizance. Technology needs to be rebuilt and revamped as science and engineering move from one paradigm to another in this decade. The question of environmental sustainability is gaining new and visionary heights in the pursuit of environmental engineering science. Zero-discharge norms are the utmost need of the hour as technology and engineering surpasses wide and versatile frontiers. The true challenge and the definite vision of environmental engineering science and chemical process engineering are crossing visionary boundaries. The science of industrial wastewater treatment has immense challenges and scientific tribulations. Scientific arena and deep scientific landscape in wastewater treatment are in the midst of vision and forbearance. The unending concerns for zero-discharge norms and frequent environmental catastrophes are changing the face of human scientific research pursuit. In such a crucial juxtaposition of scientific history and time, AOPs and their various branches need to be reenvisioned and redefined with vicious and effective challenges.

5.11   ENVIRONMENTAL SUSTAINABILITY AND THE WORLD OF WATER PURIFICATION

Environmental sustainability is a burning issue today in the furtherance of science and technology and future scientific endeavor. Global water technologies are gaining immense heights as human civilization surpasses one visionary boundary over another. In the world of water purification, science and engineering need to be reenvisioned and reenvisaged. The challenge and vision of water purification and drinking water treatment are veritably changing the scientific horizon. Sustainable development and industrial water pollution control/water treatment are veritably two opposite sides of the visionary coin. The author, in this chapter, repeatedly urges the scientific community to gear forward toward newer and innovative environmental engineering techniques whether they are conventional or nonconventional. The vision of sustainability was propounded by Dr. Gro Harlem Brundtland, former Prime Minister of Norway. The Brundtland Commission of the United Nations on March 20, 1987 defined sustainability as "sustainable development is development that meets the needs of the present without compromising the ability of future generations to meet their own needs."

5.12 SCIENTIFIC UNDERSTANDING AND SCIENTIFIC INGENUITY OF ADVANCED OXIDATION TECHNIQUES

Scientific vision and deep scientific ingenuity are the wide pillars of scientific research pursuit in AOPs and ozonation studies today. Environmental engineering science and chemical process engineering are on the wide path of scientific regeneration and scientific adjudication. Advanced oxidation techniques and ozonation procedure are the pillars of this treatise. The scientific profundity and scientific ingenuity are unparalleled as human civilization surpasses one visionary boundary over another. Technology needs to be reenshrined and revisited as scientific endeavor in environmental engineering science reaches new heights. Nonconventional environmental engineering techniques are immensely promising and pathbreaking. Science and vision of mankind need to be redefined as environment is at stake. Technology of AOPs has few answers for the global environmental crisis and needs to be revamped and reorganized with the course of scientific history, scientific vision, and time. Scientific ingenuity also needs to be restructured and reenvisioned at such a crucial juncture of human scientific endeavor.

5.13 TRADITIONAL AND NONTRADITIONAL ENVIRONMENTAL ENGINEERING TECHNIQUES

Traditional and nontraditional environmental engineering techniques are the forerunners toward a newer and visionary scientific regeneration and scientific fortitude. AOPs, integrated AOPs, and ozonation are the focal points of scientific endeavor today. The author, in this chapter repeatedly stresses on the effectivity and efficiency of AOPs and ozone oxidation, in particular. Environmental engineering science stands in the midst of deep scientific vision and forbearance. The world of science and technology is veritably challenged with the progress of scientific and academic rigor, which is of grave concern with respect to the environment. Environmental sustainability today is the need of human civilization. Industrial waste-water treatment, drinking water treatment, and water pollution control will all lead to a long and visionary way in the true emancipation and true realization of environmental sustainability in today's context. In vast cases, sustainable development is at deep stake as human civilization and scientific endeavor march forward. Traditional environmental engineering tools are far from effective yet groundbreaking. Nontraditional environmental engineering techniques thus is the need of the hour. The scientific success and the scientific vision are slowly evolving and gearing forward toward a newer visionary eon.

5.14   SCIENTIFIC VALIDATION, SCIENTIFIC VISION, AND SCIENTIFIC UNDERSTANDING OF OZONATION APPLICATIONS

Scientific validation as well as scientific understanding in the field of AOPs applications are the utmost need of the hour. Ozonation or ozone oxidation perspectives are changing the scientific and technological scenario. Scientific validation and scientific scale-up are the torchbearers toward a newer visionary era of environmental engineering science. Global water issues and global water research and development initiatives are ushering in a new era of scientific regeneration and deep scientific insight. Ozonation or ozone oxidation is witnessing drastic challenges as science and engineering is moving forward. Today, science and engineering stand in the midst of scientific vision and forbearance. Science is a huge colossus with a definite and purposeful vision of its own. The science of AOPs is evolving into newer knowledge dimensions and newer scientific eon. Scientific validation and scientific discernment are the hallmarks of an effective scientific research pursuit today. Zero-discharge norms and the concerns of environmental sustainability are urging the scientific domain to gear forward toward a new era of hope and determination.

5.15   FUTURE FRONTIERS OF OZONE OXIDATION AND AOPs

The technology and engineering science of ozone oxidation and AOPs today stand in the crossroads of scientific vision and deep scientific introspection. The challenge, the vision, and the deep scientific perspectives of industrial wastewater treatment, water pollution control, and drinking water treatment are changing the veritable scientific landscape. Environmental sustainability, the challenges of water pollution control, and the deep scientific prowess will all lead to a long and visionary way in the true emancipation of environmental engineering science. Future frontiers in ozone oxidation and AOPs are surpassing vast scientific boundaries. The technology of ozone oxidation needs to be reenvisioned and restructured with the passage of scientific history, scientific vision, and time. Environmental engineering science and environmental sustainability are the forerunners toward a newer era in scientific emancipation. Technological and scientific profundity and the immense scientific potential of AOPs are the forerunners toward a greater scientific discernment and a greater scientific realization in the field of environmental engineering. Scientific profundity, scientific fortitude, and scientific divinity are the hallmarks in research pursuit in industrial wastewater treatment. Human civilization and human scientific endeavor today stand in the midst of profound scientific pragmatism and sound doctrine. Industrial wastewater treatment encompasses both AOPs as well as ozone oxidation. The challenge and the vision of research forays in industrial pollution control are immense and far-reaching. In this chapter, the author pointedly focuses on the diverse areas of scientific pursuit in various AOPs and ozone oxidation. Technology and engineering science of nontraditional environmental engineering techniques are ushering in a new era in scientific research pursuit. In this chapter, the author rigorously points out the deep scientific vision, the vast scientific comprehension, and the scientific imagination behind industrial wastewater treatment and also drinking water treatment. Today, engineering science is a huge colossus with a definite vision and a definite will of its own. Environmental catastrophes and depletion of fossil-fuel resources are challenging the very scientific fabric of environmental and energy sustainability. Humankind's immense scientific prowess, the deep technological vision, and the vast scientific introspection will all lead to a long and visionary way in the true realization of science of sustainability. Technology is highly challenged today as human civilization marches forward. Wastewater crisis and zero-discharge norms in environmental engineering science are urging the scientific domain to gear forward toward newer scientific innovation and newer scientific instincts. This chapter gives an effective glimpse of the scientific success, the scientific genesis, and the deep scientific fortitude in the research pursuit in various AOPs and ozone oxidation, in particular. Technological revamping and scientific discernment are in a state of immense comprehension and vision. Future research trends and future frontiers in AOPs are surpassing scientific imagination and scientific boundaries. Human civilization's immense intellectual determination and the scientific girth are reenvisioning the course of scientific research pursuit in nontraditional environmental engineering tools such as AOPs. Deep scientific understanding is the necessity of research pursuit today as sustainability science and sustainability concerns are revitalizing the vast scientific fabric. This chapter willfully elucidates upon the scientific success and the scientific potential of the intricacies and barriers in successful implementation of AOPs in wastewater treatment and drinking water treatment. Mankind's immense scientific vision, deep scientific candor, and the futuristic vision will all lead to a long and visionary way in the true realization and the true advancement of science and engineering.[18,1,2]

5.16 FUTURE RESEARCH TRENDS AND FUTURE RECOMMENDATIONS

Future research trends in environmental engineering science are changing the face of global scientific paradigm. The technological vision, the scientific candor, and the deep scientific introspection are the imperatives and necessities of research endeavor today. Human civilization and human scientific endeavor are on the path toward newer scientific regeneration and vision. Environmental catastrophes, the introspection into science and engineering, and the futuristic vision in wastewater treatment will all lead to a long and visionary way in the successful realization of environmental sustainability. Research trends and research fortitude are today in an avenue of immense rejuvenation with the growing concerns for energy and environmental sustainability. Technology and engineering science has few answers for the devastations of environment, the vicious climate change, and the loss of fossil-fuel resources. Sustainable development at this crucial juncture is the scientific imperative of future research pursuit. This chapter deeply comprehends the true vision of energy and environmental sustainability and veritably opens up new windows of scientific innovation and vast scientific instincts in decades to come. The vision of Dr. Gro Harlem Brundtland, the former Prime Minister of Norway needs to be reenvisioned and reenvisaged with the passage of scientific history, scientific forbearance, and time. The vast aim and mission of this chapter is to present a critical overview of the scientific intricacies, scientific hindrances, and the vast scientific barriers in the advanced oxidation applications in wastewater treatment and water pollution control. Scientific justification and scientific adjudication are the veritable hallmarks of this well-researched treatise. Human scientific endeavor, the challenges of technological validation, and the march of engineering science will all lead to a long and visionary way in the true realization of sustainability science.

5.17 SUMMARY, CONCLUSION, AND SCIENTIFIC PERSPECTIVES

Scientific perspectives and wide scientific paradigm are today in the midst of deep introspection. The vision and the challenge need to be reorganized and reenvisaged as science moves willfully toward a newer eon. Environmental catastrophes and the global energy crisis are veritably stalling the future progress of human civilization and the future course of human scientific endeavor. Environmental and energy sustainability are today highly challenged as civilization drastically moves from one paradigm toward another. In this chapter, the author rigorously points toward the vast potential and the vast scientific ingenuity in advanced oxidation applications in industrial wastewater treatment. Technology of ozone oxidation and other AOPs today needs to be redefined and re-understood as science and engineering of environmental science move forward. Technological diversification in the vast areas of environmental engineering science needs to be envisioned with the passage of scientific history and time. Scientific perspectives in environmental and energy sustainability are slowly changing from one paradigm toward another. Sustainability development with respect to environment and water is the utmost need of the hour. In such a crucial juncture of deep scientific vision and scientific history, human scientific prowess and deep scientific determination will all lead to a long and visionary way in the true emancipation of environmental sustainability and environmental science today. The status of research trends in environmental engineering is witnessing immense revamping with the growing concerns of climate change, frequent environmental disasters, and the concerns of pollution control. The challenge of science goes beyond scientific imagination and needs to be reenvisioned and redefined. Technological validation and deep scientific motivation are the true answers for environmental engineering emancipation. AOPs are the few answers toward the scientific profundity behind industrial wastewater treatment and provision of pure drinking water. This chapter faces immense challenges and wide vision. The author, in this chapter, willfully presents the deep scientific vision, the scientific doctrine, and the immense scientific pragmatism behind application of nonconventional environmental engineering techniques. This chapter largely addresses certain issues of environmental sustainability with the sole aim of uncovering the latent scientific truth behind environmental engineering.

ACKNOWLEDGMENT

The author with deep respect wishes to acknowledge the contributions of Shri Subimal Palit, the author's late father and an eminent textile engineer who taught the author the rudiments of Chemical Engineering.

KEYWORDS

- ozone
- engineering
- wastewater
- vision
- advanced
- oxidation

REFERENCES

1. www.google.com (accessed date October 1, 2017).
2. www.wikipedia.com (accessed date October 1, 2017).

3. Agustina, T. E.; Ang, H. M.; Vareek, V. K. A Review of Synergistic Effect of Photocatalysis and Ozonation on Wastewater Treatment. *J. Photochem. Photobiol., C* 2005, *6*, 264–273.

4. Rosal, R.; Rodriguez, A.; Perdigon-Melon, J. A.; Petre, A.; Garcia-Calvo, E.; Gomez, M. J.; Aguera, A.; Fernandez-Alba, A. R. Occurrence of Emerging Pollutants in Urban Wastewater and their Removal Through Biological Treatment Followed by Ozonation. *Water Res.* **2010**, *4*(2), 578–588.

5. Palit, S. Studies on Ozone-Oxidation of Dye in a Bubble Column Reactor at Different pH and Different Oxidation-Reduction Potential. *Int. J. Environ. Sci. Develop.* **2010**, *1*(4), 341–346.

6. Guo, Y.; Yang, L.; Cheng, X.; Wang, X. The Application and Reaction Mechanism of Catalytic Ozonation in Water Treatment. *J. Environ. Anal. Toxicol.* **2012**, *2*(7), 2–6.

7. Eriksson, M. Ozone Chemistry in Aqueous Solution- Ozone Decomposition and Stabilization. Licentiate Thesis, Department of Chemistry, Royal Institute of Technology, Stockholm, Sweden, 2005.

8. Zhu, Y.; Zhang, H.; Zhang, X. Study on Catalytic Ozone Oxidation with Nano-TiO_2 Modified Membrane for Treatment of Municipal Wastewater. *J. Biomimetics Biomater. Tissue Eng.* **2013**, *18*(2), 2–5.

9. Broseus, R.; Vincent, S.; Aboulfadl, K.; Daneshvar, A.; Sauve, S.; Barbeau, B.; Prevost, M. Ozone-Oxidation of Pharmaceuticals, Endocrine Disruptors and Pesticides During Drinking Water Treatment. *Water Res.* **2009**, *43*, 4707–4717.

10. Liu, P.; Zhang, H.; Feng, Y.; Yang, F.; Zhang, J. Removal of Trace Antibiotics from Wastewater: A Systematic Study of Nanofiltration Combined with Ozone-Based Advanced Oxidation Processes. *Chem. Eng. J.* **2014**, *240*, 211–220.

11. Munter, R. Advanced Oxidation Processes- Current Status and Prospects. *Proc. Est. Acad. Sci. Chem.* **2001**, *50*(2), 59–80.

12. Sharma, S.; Ruparelia, J. P.; Patel, M. L. A General Review on Advanced Oxidation Processes for Wastewater Treatment. International Conference on Current Trends in Technology, NUICONE-2011,Institute of Technology, Nirma University: Ahmedabad, India, (8–10 December, 2011).

13. Jelonek, P.; Neczaj, E. The use of Advanced Oxidation Processes (AOP) for the Treatment of Landfill Leachate. *Inz. Ochr. Srodowiska* **2012**, *15*(2), 203–217.

14. Gilmour, C. R. Water Treatment using Advanced Oxidation Processes: Application Perspectives. Master of Engineering Science Thesis, Western University, Canada, 2012.

15. Hashim, M. A.; Mukhopadhyay, S.; Sahu, J. N.; Sengupta, B. Remediation Technologies for Heavy Metal Contaminated Groundwater. *J. Environ. Manage.* **2011**, *92*, 2355–2388.

16. Cheryan, M. *Ultrafiltration and Microfiltration Handbook;* Technomic Publishing Company, Inc.: USA, 1998.

17. Palit, S. Filtration: Frontiers of the Engineering and Science of Nanofiltration-A Far-Reaching Review. In *CRC Concise Encyclopedia of Nanotechnology (Taylor and Francis);* Ortiz-Mendez, U., Kharissova, O. V., Kharisov, B. I., Eds.; CRC Press: USA; 2016; 205–214.

18. Palit, S. Advanced Oxidation Processes, Nanofiltration, and Application of Bubble Column Reactor. In *Nanomaterials for Environmental Protection;* Kharisov, B. I., Kharissova, O. V., Rasika Dias, H. V., Eds.; Wiley: USA, 2015; 207–215.

CHAPTER 6

NOVEL APPLICATIONS OF MULTIFERROIC HETEROSTRUCTURES

ANN ROSE ABRAHAM[1], SABU THOMAS[2,3], and
NANDAKUMAR KALARIKKAL[1,2,*]

[1]*School of Pure and Applied Physics, Mahatma Gandhi University,
Kottayam, Kerala 686560, India*

[2]*International and Inter University Centre for Nanoscience and
Nanotechnology, Mahatma Gandhi University, Kottayam, Kerala
686560, India*

[3]*School of Chemical Sciences, Mahatma Gandhi University,
Kottayam, Kerala 686560, India*

Corresponding author. E-mail: nkkalarikkal@mgu.ac.in

ABSTRACT

Multiferroic heterostructures in which ferroelectric and ferromagnetic orders are coupled enable control of magnetism using electric fields. Hence, they are of increasing demand for realizing ultra-fast, compact, miniaturized, and ultra-low power spintronics. The wide demand has boosted the development of novel *multiferroic* heterostructures. In this chapter, we present an overview of recent advances in the multiferroic heterostructures, highlighting the strain tuning for the development of strong magnetoelectric coupling effects and their potential applications. The varied applications of the multiferroics in the field of memory devices, magnetic data storage, nonlinear optical applications, electro-optic responses, plasmonic waveguide modulators, magnetoelectric device applications, and so forth, are discussed in this chapter. To conclude, we present a brief overview of functionalities of these various multiferroic heterostructures that enable accomplishment of novel functional devices.

6.1 INTRODUCTION

The cross-coupling effects between electric and magnetic order parameters in multiferroics[13] contribute to the electric field control of magnetism, that is crucial for the development of emerging low-power magnetoelectric (ME) devices.[61] The prospect of manipulating the magnetic or electric structure by means of either electric or magnetic fields is made possible by the renaissance of multiferroics.[26] Great success has been made in this field by the continued efforts of the researchers to develop hybrid multiferroic structures that exhibit significant value of ME coupling at room temperatures. Manipulated hybrid structures with desired material properties are obtained through various strategies such as functionalization, interfacing layers, epitaxial growth, and other techniques.

6.2 MULTIFERROICS

Multiferroics[25] are those materials in which more than one of the *primary ferroic order* parameters such as ferromagnetism, ferroelectricity, ferroelasticity, or ferrotoridicity are coupled. Multiferroics[34,61] are scarce due to their mutually exclusive nature of ferromagnetism and ferroelectricity (d^0 vs. d^n problem).[46] In multiferroics,[16,55] ferromagnetism, the spontaneous ordering of orbital and spin magnetic moments, and ferroelectricity, the spontaneous ordering of electric dipole moments, can coexist in one material in the absence of external electric and magnetic fields. The third type of order, spontaneous deformation, which leads to ferroelasticity can also coexist. Boracites were probably the first known multiferroics. The Kittel's law is observed to be applicable for multiferroics.[35]

6.2.1 MAGNETOELECTRIC (ME) EFFECT

Multiferroic materials enjoy coexistence of two or more ferroic order parameters.[44] Ferroic materials enjoy a spontaneous electric polarization

(P_i), magnetization (M_i), or strain (ε_{ij}). The magnetic field-induced change of electric polarization is called ME effect in single-phase materials.[78] In multiferroic composites, the ME effect is recognized as piezoelectric effect induced by magnetostriction. The coexistence of spontaneous polarization and spontaneous magnetization allows coupling between polarization and magnetic field, which is called direct magnetoelectric effect,[15] or between magnetization and the electric field is known as converse magnetoelectric effect.[21,79] Such fascinating coupling effects in multiferroics can be utilized to realize additional functionalities on existing devices. The direct magnetoelectric effect may enable the transformation of a magnetic signal to an electric voltage signal with high sensitivity and hence has the potential to revolutionize the huge and expensive superconducting quantum interference devices that operate only at cryogenic temperatures.[21]

The ME effect in a single-phase crystal is explained by the Landau theory in terms of free energy F of the system.[19] The multiferroic system is fragmented into domains[20] separated by transition regions called multiferroic domain walls.[48] Nanocrystalline $BaTiO_3$ exhibit multiferroic properties.[33]

Multiferroics are divided into two classes, depending on the origin of their polarization: Type I and Type II multiferroics.[28] In Type-I multiferroics, magnetism and ferroelectricity exist independent of each other and exhibit high polarizations and high critical temperatures.[57] (1) bismuth-based compounds, such as bismuth manganite ($BiMnO_3$),[8] bismuth ferrite ($BiFeO_3$),[38] are lone pair active multiferroics. The lone pair Bi^{3+} play the role in field emission (FE).[3] (2) Frustrated $LuFe_2O_4$ is an example of charge-order-driven multiferroics.[67] (3) In geometric multiferroics, tilting or rotation of different sublattices are responsible for ferroelectric and magnetic orders. The hexagonal manganite, $YMnO_3$ is an example.

In Type-II multiferroics, ferroelectricity is induced by an exotic type of magnetic ordering and are also known as magnetic multiferroics.[28,57] FE and FM orders are strongly coupled in these systems. Type-II multiferroics are subdivided into two classes: spiral Type-II multiferroics and collinear Type-II multiferroics. In spiral Type-II multiferroics, ferroelectricity is induced by magnetic spiral structures. Spin–orbit coupling leads to ME coupling in spiral multiferroics. In collinear Type-II multiferroics, ferroelectricity is induced by collinear magnetic structures. The collinear magnetic structured manganite Ca_3CoMnO_6 is an example.[29]

6.2.2 MAGNETOCAPACITANCE AND MAGNETODIELECTRIC EFFECTS

In single-phase multiferroics, the magnetic and electric ordering occur at different temperatures (a high ferroelectric transition temperature and a low magnetic transition temperature), and hence exhibit weak coupling between ordered parameters.[58] $BiFeO_3$ (BFO)[3,12,64] is the only known multiferroic with high transition temperatures.[10] But bismuth ferrite has rather weak ME coupling coefficient, due to large leakage current caused by the presence of oxygen vacancies. Hence, it is unrealistic for practical applications.[58] Hence, search for new hybrid multiferroics with the magnetic and ferroelectric transition temperatures beyond the room temperature[18] are of great interest. Artificial multiferroic heterostructures[13] comprising ferromagnetic nanostructures or ferrites[59] are developed. Spintronic oxides of ferrites have extensive applications in nanodevices[75,77] and in the field of biomedicine.[41]

The magnetic field-induced change of capacitance is called as the magneto-capacitance (MC) effects.[73] Single-phase multiferroics show very low MC effect and must be operated under very strong magnetic fields. In multiferroic composites, MC effects arise with great probability, in materials that enjoy an isotropic character. The composite materials act as a dais for conductor-dielectric interface contacts oriented perpendicular to the direction of electric current flow.

The dielectric spectroscopic studies in the presence of external magnetic fields are used to observe the indirect ME coupling. The change in dielectric permittivity with field is called the magnetodielectric (MD) effect.[56] Several strain-mediated[7,23] multiferroic systems,[1] combining ferrites and ferroelectrics[4,6,9,11,14,22,30,31,47,50,52,54,56,64,71,76] have been investigated.

6.3 APPLICATIONS OF MULTIFERROICS

Ferroics enjoy features such as hysteric domain switching, high values of response functions (e.g., dielectric (χ_E) and magnetic (χ_M) susceptibilities), and coupling effects such as piezoelectricity (d_E) and magnetostriction (d_M). These unique features enable multiferroics for a variety of applications such as electric field sensors, low magnetic field sensors, ultrahigh-density magnetic memories,[44] actuators, microelectromechanical systems, ultralow-power tunable radio frequency/microwave ME devices, information storage,[21] ultrahigh capacity computer memory chips, magnetoelectric, and optomagnetic storage materials,[39] and so forth.

6.3.1 MAGNETIC DATA STORAGE

The coupling between magnetic and electric dipoles is very important for energy-efficient magnetic data storage.[37] Exploiting ME effect in spintronics technology is a promising way to control magnetization with E-field rather than with electric current or magnetic field in order to keep away from the dilemma of heat dissipation and stray magnetic field. Incorporation of multiferroic-based components into nanoscale applications, for example, in form of nanosized tunnel junctions with multiferroic barriers, enable additional degrees of freedom in manipulation with spin and charge in spintronic devices.[64] Spintronics technology exploits magnetic tunnel junctions[24] based on common dielectrics such as Al-oxide, MgO, and so forth. MgO magnetic tunnel junctions[37] have attracted the interest of researchers worldwide in recent years because of its very high tunnel magnetoresistance ratio and high thermal stability. The combination of electroresistance and magnetoresistance effects at a ferromagnetic–ME tunnel junction can result in four-state memory effect and help to realize power-efficient futuristic memory devices.[17,44]

6.3.1.1 THEORY: MULTIFERROICS FOR MAGNETIC DATA STORAGE

Multiferroic composites can be used for ultrahigh-density magnetic data storage beyond the superparamagnetic limit. An increase in the magnetic storage density is possible by shrinking of magnetic grain size. Vopson et al have proposed a system of magnetic grains on the piezo-substrate that forms a multiferroic composite system function as magnetic data storage medium, for areal densities beyond the superparamagnetic limit.[70] The elastomechanical coupling between the magnetic data storage layer and the piezo-ferroelectric substrate that reinforces the thermal stability of the magnetic grains, enables data write process. The Stoner–Wohlfarth magnetization reversal model (S–W formalism) was employed to theoretically study the potential of this notion. The hybrid multiferroic system of thin film of patterned magnetic grains deposited onto a dynamic piezo-substrate exhibits elastomechanical coupling which modifies the magnetic properties of the magnetic grains. Theoretically, in the S–W formalism each individual magnetic grain is assumed to be a uniaxial single domain particle and the magnetostatic intergrain interactions are neglected since the separation distance between the grains is assumed to be large enough. If φ is the angle between the applied field and the magnetic easy axis (EA), β is the angle between the applied external stress and the EA, and θ is the angle between the magnetization and the EA (see Fig. 6.1b), then the total energy W_{tot}, for a magnetic particle of volume V, under applied magnetic field and external stress σ, is given by,

$$W_{tot} = K_a V \sin^2\theta - HM_s V \cos(\varphi - \theta) - \frac{3}{2} M \lambda_s \sigma_1 V \sin^2(\beta - \theta)$$

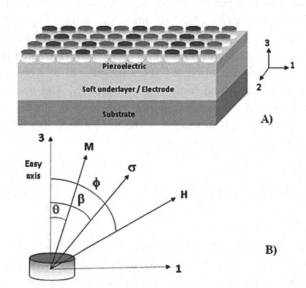

FIGURE 6.1   (See color insert.) (a) Schematic of a patterned magnetic recording medium in which magnetic bits form a strain-mediated multiferroic system with the p iezoelectric substrate; (b) Uniaxial single domain magnetic particle in applied H field and σ(r) stress induced by the piezo-substrate when electrically activated.

Source: Reprinted from ref 70 (with permission from AIP Publishing).

In the case of a zero applied stress, the anisotropy field, H_a is given by the well-known relation:

$$H_a = 2\frac{K_a}{M_S}$$

The anisotropy field, H_a is rewritten in terms of an effective anisotropy field constant (K_a^{eff}),

$$H_a = 2\frac{K_a^{eff}}{M_S}$$

The effective anisotropy field constant is inferred as

$$K_a^{eff} = \left(K_a + \frac{3\lambda_S Y d_{31} V_3}{2t_e} \right)$$

The anisotropy field is given by

$$H_a = \frac{1}{M_S}\left(2K_a + \frac{\lambda_S Y d_{31} V_3}{t_e} \right)$$

This formula explains the effective magnetocrystalline anisotropy of a multiferroic coupled magnetic particle. The anisotropy field variations are attributed to the changes in the effective anisotropy field constant of the coupled multiferroic system. The above relation proves that the effective magnetic anisotropy constant is dependent on the choice of materials chosen and the voltage polarity. It is, thus, clearly proven that the elasto-mechanical coupling alters the effective magnetocrystalline anisotropy of magnetic grains that are coupled to a piezo-electric substrate. In this case, very small grains turn out to be thermally unstable and the multiferroic-coupling scheme is exploited to increase the magnetocrystalline effective anisotropy artificially. This can be efficiently utilized to thermally stabilize superparamagnetic particles for ultrahigh-density magnetic data storage.

6.3.2 ELECTRO-OPTIC RESPONSES

Multiferroic bismuth ferrite nanoparticle-doped nematic liquid crystal (NLC) device has been reported to exhibit superior electro-optic (EO) response.[38] $BiFeO_3$ is a unique material that simultaneously exhibits ferro-electric (Curie temperature $(T_C) = 830°C$) and long-range antiferromag-netic G-type (Neel temperature $(T_N) = 370°C$) ordering which has superior electric and magnetic properties.[38] Multiferroic bismuth ferrite (BiFeO$_3$/BFO) nanoparticles (NPs) were found to display superior EO response when doped in NLC. The superior EO responses obtained in multiferroic bismuth ferrite (BiFeO3/BFO) NPs were reported by Navek et al. The superior EO response of the BFO NP–NLC system may be due to the strong and noteworthy room temperature ferroelectric polarization of magnitude 90–95 µC/cm^2 exhibited by BFO along the pseudocubic (111) direction. Bismuth ferrite NPs were doped in NLC. About 0.15 wt.% of BFO was employed for the purpose and a 5 µm-thick cell was developed. The liquid crystal device was addressed in the large signal regime by an amplitude modulated square wave signal at the frequency of 100 Hz. A total optical response time (rise time + decay time) of 2.5 ms was achieved for ~7 V$_{rms}$. The restoring force assists to diminish the decay time of the device. The fast response time achieved is contributed by the viscoelastic constant and restoring force imparted by the locally ordered LCs induced by the multiferroic NPs. This proves that multiferroics play a significant impact on improving the image quality and performance of the devices and open up new avenues for the production of modulators, displays, adaptive lenses, and other EO devices.[38]

6.3.3 OPTICAL MODULATION

Materials such as bismuth ferrite (BiFeO$_3$, BFO) or barium titanate (BaTiO$_3$, BTO), possess promising features for low loss optical modulation.[5] Under applied voltage, the ferroelectric domains can be partially reoriented from the in-plane orientation (with an ordinary refractive index n_o) to the out-of-plane orientation (with extraordinary index n_e). Thus, the refractive index for a field polarized along one axis can be changed, and control of propagating signal is achieved. Variation of the applied voltage provides a varying degree of domain switching, and thus the required level of propagating signal modulation can be realized. BTO was shown to provide high performance for photonic thin film modulators, as well as EO properties in plasmonic interferometer-based and waveguide-based modulators.

Plasmonic modulators that employ bismuth ferrite as a tunable material are used for dynamic signal switching in photonic integrated circuits. The bismuth ferrite core is sandwiched between metal plates (metal insulator–metal configuration), which also serve as electrodes. The core changes its refractive index by means of partial in-plane to the out-of-plane reorientation of ferroelectric domains in bismuth ferrite under applied voltage. As a result, guided modes change their propagation constant and absorption coefficient, allowing light modulation in both phase and amplitude control schemes. Owing to high field confinement between the metal layers, the existence of mode cut-offs for certain values of the core thickness, and near-zero material losses in bismuth ferrite, efficient modulation performance is achieved. For the phase control scheme, the π phase shift is provided by a 0.8-μm long device with propagation losses 0.29 dB/μm. For the amplitude control scheme, up to 38 dB/μm extinction ratio with 1.2 dB/μm propagation loss is predicted. However, BFO has higher birefringence with refractive index difference $\Delta n = 0.18$ nearly three times higher than in BTO. Recently, a strong change of refractive index in BFO was demonstrated and proposed for EO modulation. These plasmonic structures are beneficial for waveguiding and enhanced light matter interaction.

6.3.4 ME COUPLING EFFECTS

Enhanced MC has been reported by ferrites in combination with ferroelectrics. Pachari et al.[40] have reported the enhanced MC effect in $BaTiO_3$–ferrite composite systems. Composites of (1-X) $BaTiO_3$: X ($CoFe_2O_4$/ $ZnFe_2O_4$/$Co_{0.5}Zn_{0.5}Fe_2O_4$) (where X=20, 30, and 40 wt.%) were prepared by conventional solid-state mixing route. The existence of two different morphologies such as plate-like ($\sim$5 μm with a thickness $\sim$1 μm) and fine agglomerated spherical shape of tetragonal $BaTiO_3$ with the polyhedral morphology of cubic ferrites in the prepared composite systems. The percentage of ferrite phase is found to determine the unidirectional or random orientation of the plate-like morphology of $BaTiO_3$ in these systems. The magnetoresistance effect was analyzed using magnetoimpedance study as a function of frequency as well as Cole–Cole plot. Type and percentage of ferrites were found to highly influence the magnetoresistance effect. The MC responses robustly depend on the magnetoresistance of grain or grain boundary, the magnetostriction of the ferrite phase, and concentration of ferrite phase. The combined effect of phase morphology and magnetoresistance has led to the enhancement of MC effect in these composites. The MC values were found to be in the range between -3 and -9, -0.5 and -7, and $+1.5$ to -1.5 for $BaTiO_3$: $CoFe_2O_4$, $BaTiO_3$:

$ZnFe_2O_4$, and $BaTiO_3$: $Co_{0.5}Zn_{0.5}Fe_2O_4$ composites, respectively depending on the percentage of the ferrite phase.

Fabrication of exchange-biased artificial ferromagnetic–multiferroic core–shell nanostructures[51] were reported by Shi et al. The schematic of the exchange-biased hybrid coaxial nanostructures of ferromagnetic metal (Ni) and multiferroic bismuth ferrite ($BiFeO_3$) is shown in Figure 6.2.

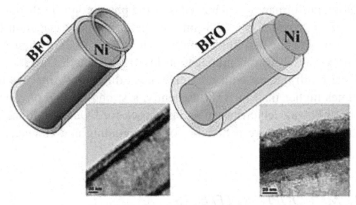

FIGURE 6.2 Schematic of exchange-biased hybrid coaxial nanostructures of ferromagnetic metal (Ni) and multiferroic bismuth ferrite ($BiFeO_3$).

Source: Reprinted with permission from ref 51. © 2014 Royal Society of Chemistry.

Exchange-biased hybrid core–shell nanostructures containing ferromagnetic metal (Ni) and multiferroic bismuth ferrite ($BiFeO_3$)[65] serve as the core and shell of the coaxial nanostructures, were fabricated by a two-step method as shown in Figure 6.3.

The ferromagnetic Ni cores are exchange coupled to the multiferroic BFO shell. The combined ferroelectric and antiferromagnetic functionalities of bismuth ferrite were utilized and an exchange bias effect was observed in the system. The scanning electron microscope (SEM) and transmission electron microscopy (TEM) images of Ni–BFO core–shell nanotubes are shown in Figure 6.4. The magnetic, ferroelectric properties and the magnetic reversal mechanism of the ferromagnetic–multiferroic core–shell nanostructures were studied. The rough core–shell interfaces are observed to reduce the exchange bias effect of the samples. Thus, ferromagnetic/ferroelectric two-phase, one-dimensional nanomaterials that allow electric field control of magnetization or magnetic field control of polarization is realized.

FIGURE 6.3 Ni–BFO core–shell nanostructures prepared through a template assisted solgel and electrodeposition two-step route.

Source: Reprinted with permission from ref 51. © 2014 Royal Society of Chemistry.

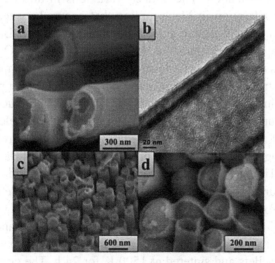

FIGURE 6.4 (a) SEM image of Ni–BFO core–shell nanotubes. (b) TEM image of a single Ni–BFO core–shell nanotube. (c and d) SEM images of Ni–BFO core–shell nanowires are grown in AAO membranes with a pore size of 300 nm.

Source: Reprinted with permission from ref 51. © 2014 Royal Society of Chemistry.

Naik and Mahendiran et al.[36] have reported the magnetic and ME studies in pure and cation-doped antiferromagnetic $BiFeO_3$. $BiFeO_3$ and $Bi_{0.7}A_{0.3}FeO_3$ (A=Sr, Ba, and $Sr_{0.5}Ba_{0.5}$) were prepared by the conventional solid-state reaction. Stoichiometric mixtures of Bi_2O_3, $SrCO_3$, $BaCO_3$, and Fe_2O_3 were mixed and grinded in an agate mortar and preheated for 5–6 h in the temperature range 800°C (x=0) − 850°C (x=0.3). Phase identification and structural characterizations were performed on the polycrystalline samples. The effect of divalent cation (A) substitution on magnetic and ME properties in $Bi_{1-x}A_xFeO_3$ (A=Sr, Ba, and $Sr_{0.5}Ba_{0.5}$; x=0 and 0.3) was observed. Magnetization exhibited by the divalent cation-doped samples increases with the size of the dopants, evidenced by distinct hysteresis loops and increase in the magnitude of spontaneous magnetization. This is due to the effective suppression of spiral spin structure. Thus, divalent cation doping enhances the magnetization in the antiferromagnetic $BiFeO_3$ It is also observed that with increasing size of the A cation, the transverse (T-α_{ME}) magnetoelectric coefficients increase in magnitude and exceeds the longitudinal (L-α_{ME}) ME coefficient. The highest transverse ME coefficient, T-α_{ME}=2.1 mV/cm·Oe is displayed by A=$Sr_{0.15}Ba_{0.15}$ in the series, despite the fact that it is not the compound with maximum saturation magnetization. The noticed variations in the ME coefficients point out the alteration in ME coupling that occurs between the phases in these compounds. The codoped compound, A=$Sr_{0.15}Ba_{0.15}$ exhibits smaller leakage current than the antiferromagnetic $BiFeO_3$. Additional studies such as magnetostriction and electrical polarization measurements at higher electric fields will throw more light on the product properties.

The introduction of impurities into BTO has a strong influence on phase transition nature and optical, magnetic, ferroelectric, and dielectric properties of $BaTiO_3$. Guo et al. have reported development and investigation on structural, magnetic, and dielectric properties of Fe-doped $BaTiO_3$ solids.[18] Fe-doped $BaTiO_3$ (BFTO) was prepared by the standard solid-state reaction and its structural, magnetic, and ferroelectric properties were investigated. A two-step process was adopted for the preparation of BFTO solids by the standard solid-state reaction technique. High purity $BaTiO_3$ and Fe_3O_4 were well-mixed, ground, and preheated for 10 h at 1200 K. The samples were pressed into pellets and sintered at 1500 K for 24 h. The crystal structure of the samples was investigated by X-ray diffraction (XRD) and Rietveld refinement analysis using General Structure Analysis System package. The BFTO samples were identified to possess tetragonal structure from the

XRD pattern. The ab-plane expansion and out-of-ab-plane shrinkage of the BFTO phases on doping were detected by the Rietveld refinements of XRD data. The decrease in the unit cell volume indicates the lattice variation caused by Fe doping. FE-SEM was used to study the microstructure of the BFTO solids. The X-ray photoelectron spectroscopy (XPS) measurements that illustrate chemical composition and valence state of ions unveils that Fe^{3+} and Fe^{4+} ions substitute Ti^{4+} ions in the crystal lattice. The transition temperature T_C is found to shift to high temperature (from 770 to 800 K) with increase in doping level (from x = 0.10 to x = 0.40). The samples display the simultaneous existence of both ferromagnetic and ferroelectric ordering at room temperature as noticed from the results of temperature-dependent dielectric studies and magnetic hysteresis studies. The doping of magnetic element Fe brings about ferromagnetic order for the samples, and the increase in magnetic moment value for each Fe atom (from 0.70 to 1.55 μB) in BFTO samples. The mechanism of origin of ferromagnetic coupling among the BFTO samples is endorsed to the double exchange interactions of $Fe^{3+}-O_2-Fe^{4+}$ ions. The ferroelectric and ferromagnetic transition temperatures were observed at about 401 K and above 770 K. The unusual change observed in the magnetization of the samples with high doping level at the ferroelectric Curie temperature reveals signatures of probable coupling between the ferromagnetic and ferroelectric orders.

Anusree et al. have reported the development of nanocomposites (NCs) of bismuth ferrite (BFO) and strontium hexaferrite (SRF) with 10–40 M% by the solgel method.[12] BFO, gives only a weak ferromagnetic hysteresis loop with low saturation magnetization, since being a G-type antiferromagnet (spiral spin structure). A composite of BFO with SRF is developed, with an aim to improve the ferroelectric, magnetic, and magnetoelectric coupling properties of BFO. Strontium hexaferrite ($SrFe_{12}O_{19}$, SRF) is a ferrimagnetic material that possesses high magnetic moment. To establish the role of SRF in enhancing magnetic and dielectric properties of BFO, structural, magnetic, dielectric, and MD properties of the BFO–SRF NCs have been studied. (1−x) $BiFeO_3$ (BFO) and (x) $SrFe_{12}O_{19}$ (SRF) (x = 0.1, 0.2, 0.3, and 0.4) NCs were prepared by solgel route. Presence of pure phases of both BFO and SRF in the NCs were confirmed by Rietveld analysis of XRD data. Morphology of the NPs is examined by TEM images and found to be spherical in shape with the average size distribution of 30 nm. The ferrimagnetic behavior of the NCs is studied by M-H measurements performed at 300 and 10 K. An increasing trend of saturation magnetization values with increasing content of SRF is observed. Dielectric studies reveal the dispersive nature of the dielectric constant (ε_r) with frequency at room temperature. With increasing temperature, ε_r–T measurements reveal an increased dielectric constant of the NCs and the anomaly exhibited at around the antiferromagnetic transition temperature of BFO, indicates the magnetoelectric coupling in the NCs. the enhancement of magnetoelectric effect in the NCs is confirmed by the MC effect. The presence of only Fe^{3+} ions in the normal crystallographic sites of BFO and SRF is demonstrated by ^{57} Fe Mössbauer spectroscopy.[60] The presence of magnetoelectric effect in the composites is investigated by the MC effect which indicates the change in dielectric constant of a material in the presence of magnetic field. MC is defined as MC $= [\varepsilon_r\,(H) - \varepsilon_r\,(0)]/\varepsilon_r\,(0)$ [ε_r (H)$-\varepsilon_r$ (0)]/ε_r (0), where $\varepsilon_r\,(H)$ and $\varepsilon_r\,(0)$ denote dielectric constants at applied field H and zero field, respectively.

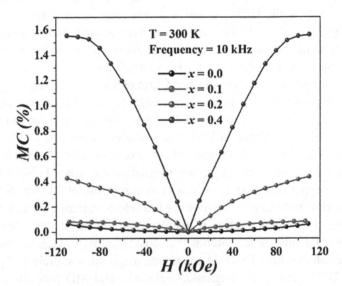

FIGURE 6.5   **(See color insert.)** Magnetic field-induced change in dielectric constant at 10 kHz of the pristine BFO and NCs (x Hz of the pristine BFO and NCs t 10ed.

Source: Reprinted from ref 12 (with permission from AIP Publishing).

The distribution of iron atoms at the available sites and their chemical environment is studied by Mössbauer spectroscopy. Mössbauer spectra of all NCs at room temperature are shown in Figure 6.6. Spectra of all NCs reveal the presence of all crystallographic inequivalent Fe $^{3+}$ ordered sites of both BFO and SRF phases confirming the formation of the composites. The improvement of magnetic as well as magnetodielectric properties of the NCs with respect to the pristine BFO is clearly observed from the Mosbauer and dielectric studies.

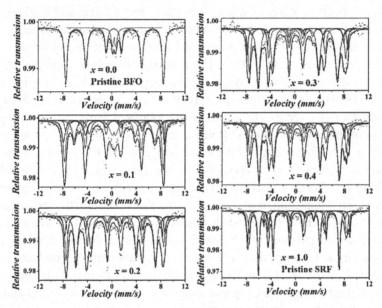

FIGURE 6.6 (See color insert.) Room temperature Mössbauer spectra of all samples (open circles: experimental, black: fitted, blue: 4f2, green: 2a and 4f1, red: 12 k, purple: 2b, magenta: BFO, dark yellow: $Bi_2Fe_4O_9$).

Source: Reprinted from ref 12 (with permission from AIP Publishing).

Sharma et al.[49] have reported the development of multiferroic particulate composites comprising $BaTiO_3$ as ferroelectric phase and $CoFe_{1.8}Zn_{0.2}O_4$ as ferrite phase and investigation of its structural, dielectric, ferromagnetic, ferroelectric, and AC conductivity attributes. $BaTiO_3$–$CoFe_{1.8}Zn_{0.2}O_4$ multiferroic composites with the formula $(1-x)$ $BaTiO_3$–(x) $CoFe_{1.8}Zn_{0.2}O_4$ (where $x = 10, 20, 30,$ and 40 wt.%) were fabricated by the solid-state reaction method. $CoFe_{1.8}Zn_{0.2}O_4$ (CZF) was prepared by using Co_3O_4, Fe_2O_3 and ZnO and $BaTiO_3$ (BT) from milling of $BaCO_3$ and TiO_2 in stoichiometric ratio. Thorough mixing of the constituent phases BT and CZF yield the multiferroic composites $(1-x)$ $BaTiO_3$–(x) $CoFe_{1.8}Zn_{0.2}O_4$. Formation of the distinct phases (single phase tetragonal pervoskite structure for BTO and cubic spinel structure for CZF) and the multiferroic composite were confirmed by XRD technique. The surface morphology and increase in grain size with increasing ferrite concentration are revealed by SEM images. The effect of ferrite content on the dielectric, electric, and magnetic properties of the composites is also studied. Dielectric constant increases as ferrite fraction increase, due to the space charge effect and the hopping conduction mechanism. The ferroelectric to the paraelectric phase transition of the $BaTiO_3$ phase is indicated by the peak observed in the low-temperature range (120–150°C). The conduction mechanism in the NCs is understood from the AC conductivity and explained by the jump relaxation model (JRM). The ferroelectric nature of the composite is studied from the polarization (P) versus electric field (E) loop and ferromagnetic properties from the magnetization (M) versus magnetic field (H) loops recorded using a vibrating sample magnetometer. For all the composites, the dielectric dispersion is observed at low frequency which can be explained by Maxwell–Wagner interfacial polarization. With the increase in ferrite content, an enhancement in dielectric constant is observed. The apparent high values of dielectric constant observed are explained by the effect of space charges and hopping conduction mechanism that have an imperative part in such composites. The nature of variation of AC conductivity is explained by the JRM. Further, all the composites display P–E and M–H loops at room temperature representing the ferroelectric and ferromagnetic nature of the samples at the same time.

Mapping of strain-mediated ME phenomena[2] in the barium titanate–barium hexaferrite ($BaTiO_3$–$BaFe_{12}O_{19}$) composite system,[62] using high-resolution techniques including switching spectroscopy piezoresponse force microscopy and spatially resolved confocal Raman microscopy has been reported. It is established that magnetic field affects the polarization switching processes such as domain nucleation and propagation and hence both the local piezoelectric coefficient and polarization switching parameters were observed to change on the application of an external magnetic field. This is attributed to the ME-induced modulation of the depolarization field.

MD coupling is reported to be improved by surface functionalization of nickel NPs in Ni and polyvinylidene fluoride nanohybrids.[32] The fabrication of magnetic microstructures with multifarious two-dimensional geometric shapes employing the magnetically assembled iron oxide (Fe_3O_4)

and cobalt ferrite ($CoFe_2O_4$) NPs by the magnetic assembly method were reported by Velez et al.[68] Surface-effect-enhanced ME coupling in FePt/PMN-PT multiferroic film heterostructures[74] were reported. The FePt films with different thickness, deposited on Pb ($Mg_{1/3}Nb_{2/3}$) O_3–$PbTiO_3$ (PMN-PT) substrates yield a symmetric "butterfly" shaped $\Delta M/M$-E_{dc} loops through thestrain-mediated magnetoelectric coupling. A self-assembled multiferroic nanostructure, composed of $PbTiO_3$ (PTO) pillars embedded in a $CoFe_2O_4$ (CFO) matrix has been reported by Tsai et al.[63] The epitaxial multiferroic $PbTiO_3$–$CoFe_2O_4$ nanostructures were deposited on MgO (001) by pulsed laser deposition.

6.3.5 MEMORY APPLICATIONS

An exciting application of multiferroic bits is to store information in the magnetization M and the polarization P states. Such a four-stage memory (two magnetic $M\uparrow\downarrow$ and two ferroelectric $P\uparrow\downarrow$) is feasible by exploitation of multiferroics. ME coupling enables nonvolatile memory device applications, where information is written magnetically and stored in the electric polarization. Multiferroics bits might also help to increase the magnetic anisotropy to increase the decay time for magnetic storage.

A multiferroic-based selector-free resistive switching (RS) memory cell that displays high resistance ratio and nonlinearity factor was reported to be developed.[27] The $BiFeO_3$ nanoisland-based memory cell can perform as a simple and competent building block of high-density resistive random access memory (ReRAM). Highly nonlinear and bistable current–voltage (I–V) characteristics are necessary in order to realize high-density ReRAM devices that are compatible with cross-point stack structures. The bipolar RS behaviors of nanocrystalline $BiFeO_3$ (BFO) nanoislands grown on Nb-doped $SrTiO_3$ substrates, with large ON/OFF ratio of 4420 were reported by Jeon et al. The BFO nanoislands exhibit asymmetric I–V characteristics with high nonlinearity factor of 1100 in a low-resistance state. Such selector-free RS behaviors are enabled by the mosaic structures and pinned downward ferroelectric polarization in the BFO nanoislands. The high value of resistance ratio and the nonlinearity factor promote the extension of BFO nanoislands to an $N \times N$ array of $N=3740$ corresponding to $\sim 10^7$ bits. Therefore, the BFO nanoisland that shows both high-resistance ratio and nonlinearity factor offers an unchallenging and highly hopeful building block of high-density ReRAM.

6.3.6   FLUORESCENCE EFFECTS

ME and fluorescence effects observed in multiferroic xBaTiO$_3$-(1-x) ZnFe$_2$O$_4$ (BTZF) nanostructures were reported.[9] The BTZF nanostructures were synthesized by solgel method using PVA as surfactant using barium acetate, tetra-n-butyl orthotitanate, zinc acetate, and ferric chloride with desired molar concentration. The precursor solution was added in PVA solution in the molar ratio of M: PVA: 5:2. The solution was dried at 250°C and annealed at 700°C for 2 h to crystallize. Different types of nanostructural shape and size have been obtained by the effect of ionic radii, surface energy, and PVA which enhances the ME/dielectric interaction between BT/ZF phases. The crystallinity of the composites was confirmed by XRD, and TEM. The improvement in magnetization of BTZF depends upon size and shape of the nanostructure, stoichiometric ratio, and occupation of cations at octahedral and tetrahedral sites. The XPS analyzes the chemical states of Fe in BTZF. The ferroelectricity is explained by nanosize effect, one-dimensional nanostructure shape, lattice distortion, and epitaxial strain between two phases. The ME coefficient and the MC is measured and explained on the basis of Maxwell–Wagner space charge and magnetoresistance. The theoretical investigation proves that the enhancement is reliant on the size/shape of the nanostructure and also the strain-induced phase transition. The BTZF nanostructures exhibit photoemission and are observed by fluorescence spectra.

Multiferroic nanostructures of Ni$_{0.6}$Zn$_{0.4}$Fe$_2$O$_4$-BaTiO$_3$ (NZF/BT)[69] prepared by two different synthesis routes, such as chemical combustion (CNZF/BT) and hydrothermal (HNZF/BT) methods were reported by Verma et al. The CNZF/BT NPs obtained by annealing at 500°C were of average size 4 nm. The NZF/BT nanostructures exhibited optical activity marked by fluorescence spectra at room temperature. The point defects, oxygen vacancies, and interstitials are primary factors that are responsible for emission spectrum of NZF/BT multiferroic nanostructure. The first blue band observed in case of NZF ferrite is due to interstitial and defect emission, while the second band is due to their phase formation. The HNZF/BT composite was observed to show the astonishingly high intensity of blue, green, yellow, and red band emissions than exhibited by CNZF/BT.

6.3.7 NONLINEAR OPTICAL APPLICATIONS

Tesfa et al have reported the nonlinear optical properties[45] of (1−x) $CaFe_2O_4$−xBaTiO$_3$ composites.[42] NCs of (1−x) $CaFe_2O_4$−xBaTiO$_3$ (x=0.1, 0.2, 0.3, 0.5, 0.7, and 0.91) were synthesized by combining solution processing and the solid-state reaction methods, respectively. The structural analysis by XRD technique and the TEM and high-resolution TEM images confirmed the formation of ferrite–ferroelectric phases and the interface of the two crystal orientations. The nonlinear optical properties of the samples were investigated by employing the single-beam open aperture Z-scan technique.[53] The obtained nonlinearity was found to fit to two-photon absorption process. The nonlinear optical parameters[66] indicate that all samples are efficient optical limiters, and the nonlinear limiting threshold values were found to increase with an increase in the ferrite content, with the highest value of 1.96 J/cm^2 for sample CaF–BT1. The improved energy band gap has triggered an enhancement in ESA and free carrier absorption has led to an increase of nonlinear optical property of the composite. The optical limiting values of the samples indicate the potential of the samples for photonic device applications.

BaTiO$_3$ NPs[72] exhibit remarkable magnetic and nonlinear optical properties.[43] The observed magnetism in BaTiO$_3$ samples processed at higher temperatures (1000°C) was due to the charge transfer effects. The samples processed at lower temperatures such as 650⁻800°C possess band gap in the range of 2.53–3.2 eV. The carrier densities in these particles were estimated to be $\sim 10^{19}$–10^{20}/cm^3 range. The magnetization is observed to increase with band gap narrowing. The higher bandgap narrowed particles exhibited increased magnetization with a higher carrier density of 1.23×10^{20}/cm^3 near to the Mott critical density. These hint the exchange of interactions between the carriers that play a dominant role in deciding the magnetic properties of these particles. The increase in charge carrier density in this undoped BaTiO3 is because of oxygen defects only. The oxygen vacancy will introduce electrons in the system and hence more charge carriers mean more oxygen defects in the system and increase the exchange interactions between Ti^{3+}, Ti^{4+}, hence high magnetic moment. The coercivity is increased from 23 to 31 nm and then decreased again for the higher particle size of 54 nm. These particles do not show photoluminescence property and hence it hints the absence of uniformly distributed distorted $[TiO_5]$–$[TiO_6]$ clusters formation and charge transfer between them. These charge transfer effects are vital in explaining the observed magnetism in high-temperature processed samples. Thus, the variation of magnetic properties such as magnetization, coercivity with band gap narrowing, particle size, and charge carrier density reveals the superparamagnetic nature of $BaTiO_3$ NPs. The nonlinear optical coefficients extracted from Z-scan studies suggest that these are potential candidates for optical imaging and signal processing applications.

Multiferroics are promising for other applications including magnetically field-tuned capacitors that tune the frequency dependence of electronic circuits with magnetic fields, and also multiferroic sensors, which through zero-field current measurements, measure magnetic fields.[29]

6.4 CONCLUSION

Multiferroic heterostructures present high potential for low-energy-consuming spintronics devices. With the mounting demand for miniaturized and low power consuming advanced data storage devices, there has been a surge of interest and revival in the field of multiferroics, due to its awesome and unique features. The varied applications of multiferroics are highlighted in the chapter. The present-day accomplishments toward energy-efficient spintronics involving multiferroic heterostructures have been reviewed. Multiferroics offer amazing potential for magnetic data storage areal density beyond the superparamagnetic size limit. The FE/MF tunnel junctions open new avenues for development of innovative functional devices. The versatile and diverse applications of these nanoarchitectures in magnetic data storage, plasmonics, ultrahigh-density magnetic memories, optical modulation, nonlinear optical limiting applications, and so forth are discussed in the chapter.

KEYWORDS

- **multiferroics**
- **magneto-electric coupling**
- **memory devices**
- **spintroncs**

REFERENCES

1. Abraham, A. R.; Raneesh, B.; Das, D.; Kalarikkal, N. Magnetic Response of Super-paramagnetic Multiferroic Core-Shell Nanostructures. *AIP Conf. Proc.* **2016**, *1731*(1), 50151. DOI: 10.1063/1.4947805.

2. Abraham, A. R.; Raneesh, B.; Woldu, T.; Aškrabić, S.; Lazovic, S.; Dohčević-Mitrović, Z. D.; Kalarikkal, N. Realization of Enhanced Magnetoelectric Coupling and Raman Spectroscopic Signatures in 0–0 Type Hybrid Multiferroic Core-Shell Geometric Nano-structures. *J. Phys. Chem. C* **2017**, acs.jpcc.6b12461. DOI: 10.1021/acs.jpcc.6b12461.

3. Andrzejewski, B.; Molak, A.; Hilczer, B.; Budziak, A.; Bujakiewicz-Kororiska, R. Field Induced Changes in Cycloidal Spin Ordering and Coincidence Between Magnetic and Electric Anomalies in BiFeO3 Multiferroic. *J. Magn. Magn. Mater.* **2013**, *342*, 17–26. DOI: 10.1016/j.jmmm.2013.04.059.

4. Anithakumari, P.; Mandal, B. P.; Abdelhamid, E.; Naik, R.; Tyagi, A. K. Enhancement of Dielectric, Ferroelectric and Magneto-Dielectric Properties in PVDF–BaFe$_{12}$O$_{19}$ Composites: A Step Towards Miniaturized Electronic Devices. *RSC Adv.* **2016**, *6*(19), 16073–16080. DOI: 10.1039/C5RA27023E.

5. Babicheva, V. E.; Zhukovsky, S. V; Lavrinenko, A. V. Bismuth Ferrite as Low-Loss Switchable Material for Plasmonic Waveguide Modulator. *Optics Express* **2014**, *22*(23), 28890–28897. DOI: 10.1364/OE.22.028890.

6. Bammannavar, B. K.; Naik, L. R. Electrical Properties and Magnetoelectric Effect in (x) Ni0.5Zn0.5Fe2O4 + (1-x) BPZT Composites. *Smart Mater. Struct.* **2009**, *18*(8), Article id 85013. DOI: 10.1088/0964-1726/18/8/085013.

7. Barone, P.; Picozzi, S. Mechanisms and Origin of Multiferroicity. *C. R. Phys.* **2015**, *16*(2), 143–152. DOI: 10.1016/j.crhy.2015.01.009.

8. Branković, Z.; Stanojević, Z. M.; Mančić, L.; Vukotić, V.; Bernik, S.; Branković, G. Multiferroic Bismuth Manganite Prepared by Mechanochemical Synthesis. *J. Eur. Ceram. Soc.* **2010**, *30*(2), 277–281. DOI: 10.1016/j.jeurceramsoc.2009.06.030.

9. Chand Verma, K.; Tripathi, S. K.; Kotnala, R. K. Magneto-Electric/Dielectric and Fluorescence Effects in Multiferroic xBaTiO$_3$–(1 − x) ZnFe$_2$O$_4$ Nanostructures. *RSC Adv.* **2014**, *4*(104), 60234–60242. DOI: 10.1039/C4RA09625H.

10. Chaturvedi, S.; Sarkar, I.; Shirolkar, M. M.; Jeng, U. S.; Yeh, Y. Q.; Rajendra, R.; Kulkarni, S. Probing Bismuth Ferrite Nanoparticles by Hard x-Ray Photoemission: Anomalous Occurrence of Metallic Bismuth. *Appl. Phys. Lett.* **2014**, *105*(10), 12–17. DOI: 10.1063/1.4895672.

11. Correas, C.; Hungría, T.; Castro, A. Mechanosynthesis of the Whole xBiFeO3–(1 −x) PbTiO3 Multiferroic System: Structural Characterization and Study of Phase Transitions. *J. Mater. Chem.* **2011**, *21*(9), 3125. DOI: 10.1039/c0jm03185b.

12. Das, A.; Chatterjee, S.; Bandyopadhyay, S.; Das, D. Enhanced Magnetoelectric Properties of BiFeO3 on Formation of BiFeO3/SrFe12O19 Nanocomposites. *J. Appl. Phys.* **2016**, *119*(23), 234102. DOI: 10.1063/1.4954075.

13. Fernandes Vaz, C. A.; Staub, U. Artificial Multiferroic Heterostructures. *J. Mater. Chem. C* **2013**, *1*(41), 6731. DOI: 10.1039/c3tc31428f.

14. Feteira, A.; Sinclair, D. C. The Influence of Nanometric Phase Separation on the Dielectric and Magnetic Properties of (1 − x) BaTiO3–xLaYbO3 (0≤ x ≤0.60) Ceramics. *J. Mater. Chem.* **2009**, *19*(3), 356. DOI: 10.1039/b816039b.

15. Fina, I.; Dix, N.; Rebled, J. M.; Gemeiner, P.; Martí, X.; Peiró, F.; Fontcuberta, J. The Direct Magnetoelectric Effect in Ferroelectric-Ferromagnetic Epitaxial Heterostructures. *Nanoscale* **2013**, *5*, 8037–8044. DOI: 10.1039/c3nr01011b.

16. Garcia, V.; Bibes, M.; Barthélémy, A. Artificial Multiferroic Heterostructures for an Electric Control of Magnetic Properties. *C. R. Phys.* **2015**, *16*(2), 168–181. DOI: 10.1016/j.crhy.2015.01.007.

17. Guo, R.; You, L.; Zhou, Y.; Lim, Z. S.; Zou, X.; Chen, L.; Wang, J. Non-Volatile Memory Based on the Ferroelectric Photovoltaic Effect. *Nat. Communications*, **2013**, *4*, 1–5. DOI: 10.1038/ncomms2990.

18. Guo, Z.; Yang, L.; Qiu, H.; Zhan, X.; Yin, J.; Cao, L. Structural, Magnetic And Dielectric Properties of Fe-Doped Batio $_3$ Solids. *Mod. Phys. Lett. B* **2012**, *26*(09), 1250056. DOI: 10.1142/S021798491250056X.

19. Hajra, P.; Maiti, R.; Chakravorty, D. Nanostructured Multiferroics. *Trans. Ind. Ceram. Soc.* **2011**, *70*, 53–64.

20. Hoffmann, T.; Thielen, P.; Becker, P.; Bohatý, L.; Fiebig, M. Time-Resolved Imaging of Magnetoelectric Switching in Multiferroic MnWO4. *Phys. Rev. B Condens. Matter Mater. Phys.* **2011**, *84*(18), 1–6. DOI: 10.1103/PhysRevB.84.184404.

21. Hu, J. M.; Chen, L. Q.; Nan, C. W. Multiferroic Heterostructures Integrating Ferroelectric and Magnetic Materials. *Adv. Mat.* **2016**, *28*(1), 15–39. DOI: 10.1002/adma.201502824.

22. Hu, P.; Kang, H.; Chen, J.; Deng, J.; Xing, X. Magnetic Enhancement and Low Thermal Expansion of (1−x−y) PbTiO3-xBi (Ni1/2Ti1/2) O3-yBiFeO3. *J. Mater. Chem.* **2011**, *21*(40), 16205. DOI: 10.1039/c1jm12410b.

23. Hu, Z.; Sun, N. X. Epitaxial Multiferroic Heterostructures. *Compos. Magnetoelectr.* **2015**, 87–101. DOI: 10.1016/B978-1-78242-254-9.00005-6.

24. Huang, W.; Yang, S.; Li, X. Multiferroic Heterostructures and Tunneling Junctions. *J. Materiomics* **2015**, *22*. DOI: 10.1016/j.jmat.2015.08.002.

25. Hur, N.; Park, S.; Sharma, P. A.; Ahn, J. S.; Guha, S.; Cheong, S.-W. Electric Polarization Reversal and Memory in a Multiferroic Material Induced by Magnetic Fields. *Nature* **2004**, *429*, 392–395. DOI: 10.1038/nature02572.

26. Narayanan, T. N.; Mandal, B. P.; Tyagi, A. K.; Kumarasiri, A.; Zhan, X.; Hahm, M. G.; Ajayan, P. M. Hybrid Multiferroic Nanostructure with Magnetic–Dielectric Coupling. *Nano Lett.* **2012**, *12*(6), 3025–3030.

27. Jeon, J. H.; Joo, H.-Y.; Kim, Y.-M.; Lee, D. H.; Kim, J.-S.; Kim, Y. S.; Park, B. H. Selector-Free Resistive Switching Memory Cell Based on BiFeO3 Nano-Island Showing High Resistance Ratio and Nonlinearity Factor. *Sci. Rep.* **2016**, *6*, 23299. DOI: 10.1038/srep23299.

28. Khomskii, D. Classifying multiferroics: Mechanisms and Effects. *Physics* **2009**, *2*. DOI: 10.1103/Physics.2.20.

29. Kreisel, J.; Kenzelmann, M. Multiferroics - the Challenge of Coupling Magnetism and Ferroelectricity. *Europhys. News* **2009**, *40*(5), 17–20. DOI: 10.1051/epn/2009702.

30. Kundu, A. K.; Pralong, V.; Caignaert, V.; Rao, C. N. R.; Raveau, B. Enhancement of Ferromagnetism by Co and Ni Substitution in the Perovskite LaBiMn 2 O 6+ δ. *J. Mater. Chem.* **2007**, *17*(31), 3347–3353.

31. Liu, R.; Zhao, Y.; Huang, R.; Zhao, Y.; Zhou, H. Multiferroic Ferrite/Perovskite Oxide Core/Shell Nanostructures. *J. Mater. Chem.* **2010**, *20*(47), 10665. DOI: 10.1039/c0jm02602f.

32. Mandal, B. P.; Vasundhara, K.; Abdelhamid, E.; Lawes, G.; Salunke, H. G.; Tyagi, A. K. Improvement of Magnetodielectric Coupling by Surface Functionalization of Nickel Nanoparticles in Ni and Polyvinylidene Fluoride Nanohybrids. *J. Phys. Chem. C* **2014**, *118*(36), 20819–20825. DOI: 10.1021/jp5065787.

33. Mangalam, R. V. K.; Ray, N.; Waghmare, U. V.; Sundaresan, A.; Rao, C. N. R. Multiferroic Properties of Nanocrystalline BaTiO3. *Solid State Commun.* **2009**, *149*(1), 1–5.

34. Martin, L. W.; Ramesh, R. Multiferroic and Magnetoelectric Heterostructures. *Acta Mater.* **2012**, *60*, 2449–2470. DOI: 10.1016/j.actamat.2011.12.024.

35. Matzen, S.; Fusil, S. Domains and Domain Walls in Multiferroics. *C. R. Phys.* **2015**, *16*(2), 227–240. DOI: 10.1016/j.crhy.2015.01.013.

36. Naik, V. B.; Mahendiran, R. Magnetic and Magnetoelectric Studies in Pure and Cation Doped BiFeO3. *Solid State Commun.* **2009**, *149*(19–20), 754–758. DOI: 10.1016/j.ssc.2009.03.003.

37. Naik, V. B.; Meng, H.; Xiao, J. X.; Liu, R. S.; Kumar, A., Zeng, K. Y.; Yap, S. Electric-Field-Induced Strain-Mediated Magnetoelectric Effect in CoFeB-MgO Magnetic Tunnel Junctions. **2013**. Retrieved from http://arxiv.org/abs/1311.3794.

38. Nayek, P.; Li, G. Superior Electro-Optic Response in Multiferroic Bismuth Ferrite Nanoparticle Doped Nematic Liquid Crystal Device. *Sci. Rep.* **2015**, *5*, 10845. DOI: 10.1038/srep10845.

39. Nguyen, T. H. L.; Laffont, L.; Capsal, J. F.; Cottinet, P. J.; Lonjon, A.; Dantras, E.; Lacabanne, C. Magnetoelectric Properties of Nickel Nanowires-P (VDF-TrFE) Composites. *Mater. Chem. Phys.* **2015**, *153*, 195–201. DOI: 10.1016/j.matchemphys.2015.01.003.

40. Pachari, S.; Pratihar, S. K.; Nayak, B. B. Enhanced Magneto-Capacitance Response in $BaTiO_3$–Ferrite Composite Systems. *RSC Adv.* **2015**, *5*(128), 105609–105617. DOI: 10.1039/C5RA16742F.

41. Pankhurst, Q. A.; Connolly, J.; Jones, S. K.; Dobson, J. Applications of Magnetic Nanoparticles in Biomedicine. *J. Phys. D Appl. Phys.* **2003**, *36*(13), 167–181. DOI: 10.1088/0022-3727/36/13/201.

42. Philip, R. Nonlinear Optical Properties of (1-x) CaFe2O4-xBaTiO3 Composites. *Ceram. Int.* **2016**, *42*(9), 11093–11098. DOI: 10.1016/j.ceramint.2016.04.009.

43. Ramakanth, S.; Hamad, S.; Rao, S. V.; Raju, K. C. J. Magnetic and Nonlinear Optical Properties of BaTiO3 Nanoparticles Magnetic and Nonlinear Optical Properties of BaTiO 3 nanoparticles. *AIP Adv.* **2015**, *57139*, 0–11. DOI: 10.1063/1.4921480.

44. Ramírez-Camacho, M. C.; Sánchez-Valdés, C. F.; Gervacio-Arciniega, J. J.; Font, R.; Ostos, C.; Bueno-Baques, D.; Raymond-Herrera, O. Room Temperature Ferromagnetism and Ferroelectricity in Strained Multiferroic BiFeO3 Thin Films on La0.7Sr0.3MnO3/SiO2/Si Substrates. *Acta Mater.* **2017**, *128*, 451–464. DOI: 10.1016/j.actamat.2017.02.030.

45. Raneesh, B.; Nandakumar, K.; Saha, A.; Das, D.; Soumya, H.; Philip, J.; Philip, R. Composition-Structure–Physical Property Relationship and Nonlinear Optical Properties of Multiferroic Hexagonal $ErMn_{1-x} Cr_x O_3$ Nanoparticles. *RSC Adv.* **2015**, *5*(17), 12480–12487. DOI: 10.1039/C4RA12622J.

46. Rao, C. N. R.; Rayan, C. New Routes to Multiferroics. *J. Mater. Chem.* **2007**, *17*, 4931–4938. DOI: 10.1039/b709126e.

47. Schileo, G.; Pascual-Gonzalez, C.; Alguero, M.; Reaney, I. M.; Postolache, P.; Mitoseriu, L.; Feteira, A. Yttrium Iron Garnet/Barium Titanate Multiferroic Composites. *J. Am. Ceram. Soc.* **2016**, *99*(5), 1609–1614. DOI: 10.1111/jace.14131.

48. Scott, J. F.; Evans, D. M.; Gregg, J. M.; Gruverman, A. Hydrodynamics of Domain Walls in Ferroelectrics and Multiferroics: Impact on Memory Devices. *Appl. Phys. Lett.* **2016**, *109*(4). DOI: 10.1063/1.4959996.

49. Sharma, R.; Pahuja, P.; Tandon, R. P. Structural, Dielectric, Ferromagnetic, Ferroelectric and ac Conductivity Studies of the BaTiO 3 – CoFe 1. 8 Zn 0. 2 O 4 Multiferroic Particulate Composites. *Ceram. Int.* **2014**, *40*(7), 9027–9036. DOI: 10.1016/j. ceramint.2014.01.115.

50. Sheikh, A. D.; Mathe, V. L. Effect of the Piezomagnetic NiFe 2 O 4 Phase on the Piezoelectric Pb (Mg 1/3 Nb 2/3) 0.67 Ti 0.33 O 3 Phase in Magnetoelectric Composites. *Smart Mater. Struct.* **2009**, *18*(6), 65014. DOI: 10.1088/0964–1726/18/6/065,014.

51. Shi, D.-W.; Javed, K.; Ali, S. S.; Chen, J.-Y.; Li, P.-S.; Zhao, Y.-G.; Han, X.-F. Exchange-Biased Hybrid Ferromagnetic–Multiferroic Core–Shell Nanostructures. *Nanoscale* **2014**, *6*(13), 7215. DOI: 10.1039/c4nr00393d.

52. Sone, K.; Sekiguchi, S.; Naganuma, H.; Miyazaki, T.; Nakajima, T.; Okamura, S. Magnetic Properties of CoFe2O4 Nanoparticles Distributed in a Multiferroic BiFeO3 Matrix. *J. Appl. Phys.* **2012**, *111*(12), 124101. DOI: 10.1063/1.4729831.

53. Sreekanth, P.; Sridharan, K.; Rose Abraham, A.; Janardhanan, H. P.; Kalarikkal, N.; Philip, R. Nonlinear Transmittance and Optical Power Limiting in Magnesium Ferrite Nanoparticles: Effects of Laser Pulsewidth and Particle Size. *RSC Adv.* **2016**. DOI: 10.1039/C6RA15788B.

54. Sreenivasulu, G.; Popov, M.; Chavez, F. A.; Hamilton, S. L.; Lehto, P. R.; Srinivasan, G. Controlled Self-Assembly of Multiferroic Core-Shell Nanoparticles Exhibiting Strong Magneto-Electric Effects. *Appl. Phys. Lett.* **2014**, *104*(5), 52901. DOI: 10.1063/1.4863690.

55. Stephanovich, V. A.; Laguta, V. V. Transversal Spin Freezing and Re-Entrant Spin Glass Phases in Chemically Disordered Fe-Containing Perovskite Multiferroics. *Phys. Chemistry Chem. Phys.* **2016**, *18*(10), 7229–7234. DOI: 10.1039/c6cp00054a.

56. Stingaciu, M.; Reuvekamp, P. G.; Tai, C.-W.; Kremer, R. K.; Johnsson, M. The Magnetodielectric Effect in BaTiO3–SrFe12O19 Nanocomposites. *J. Mater. Chem. C* **2014**, *2*(2), 325. DOI: 10.1039/c3tc31737d.

57. Su, J.; Yang, Z. Z.; Lu, X. M.; Zhang, J. T.; Gu, L.; Lu, C. J.; Zhu, J. S. Magnetism-Driven Ferroelectricity in Double Perovskite Y 2 NiMnO 6. *ACS Appl. Mater. Interfaces* **2015**, *7*, 13260. DOI: 10.1021/acsami.5b00911.

58. Sundararaj, A.; Chandrasekaran, G.; Therese, H. A.; Annamalai, K. Room Temperature Magnetoelectric Coupling in BaTi1-xCrxO3 Multiferroic Thin Films. *J. Appl. Phys.* **2016**, *119*(2), 1–7. DOI: 10.1063/1.4939068.

59. Thankachan, R. M.; Cyriac, J.; Raneesh, B.; Kalarikkal, N.; Sanyal, D.; Nambissan, P. M. G. Cr^{3+} -Substitution Induced Structural Reconfigurations in the Nanocrystalline Spinel Compound ZnFe$_2$ O$_4$ as Revealed from X-Ray Diffraction, Positron Annihilation and Mössbauer Spectroscopic Studies. *RSC Adv.* **2015**, *5*(80), 64966–64975. DOI: 10.1039/C5RA04516A.

60. Thomas, J. J.; Kalarikkal, N. Mossbauer Study of Ni, Ni-Co, and Co Ferrite Nanoparticles. *AIP Conf. Proc.* **2011**, *1349* (PART A), 1175–1176. DOI: 10.1063/1.3606283.

61. Trassin, M. Low Energy Consumption Spintronics Using Multiferroic Heterostructures. *J. Phys. Condens. Matter* **2016**, *28*(3), 33001. DOI: 10.1088/0953–8984/28/3/033001.

62. Trivedi, H.; Shvartsman, V. V, Lupascu, D. C.; Medeiros, M. S. A, Pullar, R. C.; Kholkin, A. L.; Shur, V. Y. Local Manifestations of a Static Magnetoelectric Effect in Nanostructured BaTiO3-BaFe12O9 Composite Multiferroics. *Nanoscale* **2015**, *7*(10), 4489–4496. DOI: 10.1039/c4nr05657d.

63. Tsai, C. Y.; Chen, H. R.; Chang, F. C.; Tsai, W. C.; Cheng, H. M.; Chu, Y. H.; Hsieh, W. F. Stress-Mediated Magnetic Anisotropy and Magnetoelastic Coupling in Epitaxial Multiferroic PbTiO3-CoFe2O4 Nanostructures. *Appl. Phys. Lett.* **2013**, *102*(13), 0–5. DOI: 10.1063/1.4800069.

64. Tuboltsev, V.; Savin, A.; Sakamoto, W.; Hieno, A.; Yogo, T.; Räisänen, J. Nanomagnetism in Nanocrystalline Multiferroic Bismuth Ferrite Lead Titanate Films. *J. Nanopart. Res.* **2011**, *13*(11), 5603–5613. DOI: 10.1007/s11051-010-0134-9.

65. Ummer, R. P.; B, R.; Thevenot, C.; Rouxel, D.; Thomas, S.; Kalarikkal, N. Electric, Magnetic, Piezoelectric and Magnetoelectric Studies of Phase Pure ($BiFeO_3$–$NaNbO_3$)–(P (VDF-TrFE)) Nanocomposite Films Prepared by Spin Coating. *RSC Adv.* **2016**, *6*(33), 28069–28080. DOI: 10.1039/C5RA24602D.

66. Ummer, R. P.; Sreekanth, P.; Raneesh, B.; Philip, R.; Rouxel, D.; Thomas, S.; Kalarikkal, N. Electric, Magnetic and Optical Limiting (Short Pulse and Ultrafast) Studies in Phase Pure $(1 − x) BiFeO_3$–$xNaNbO_3$ Multiferroic Nanocomposite Synthesized by the Pechini Method. *RSC Adv.* **2015**, *5*(82), 67157–67164. DOI: 10.1039/C5RA10422J.

67. van den Brink, J.; Khomskii, D. I. Multiferroicity Due to Charge Ordering. *J. Phys. Condens. Matter* **2008**, *20*(43), 434217. DOI: 10.1088/0953–8984/20/43/434217.

68. Velez, C.; Torres-D? az, I.; Maldonado-Camargo, L.; Rinaldi, C.; Arnold, D. P. Magnetic Assembly and Cross-Linking of Nanoparticles for Releasable Magnetic Microstructures. *ACS Nano* **2015**, *9*(10), 10165–10172. DOI: 10.1021/acsnano.5b03783.

69. Verma, K. C.; Singh, S.; Tripathi, S. K.; Kotnala, R. K. Multiferroic $Ni_{0.6} Zn_{0.4} Fe_2 O_4$-$BaTiO_3$ Nanostructures: Magnetoelectric Coupling, Dielectric, and Fluorescence. *J. Appl. Phys.* **2014**, *116*(12), 124103. DOI: 10.1063/1.4896118.

70. Vopson, M. M.; Zemaityte, E.; Spreitzer, M.; Namvar, E. Multiferroic Composites for Magnetic Data Storage Beyond the Super-Paramagnetic Limit. *J. Appl. Phys.* **2014**, *116*(11). DOI: 10.1063/1.4896129.

71. Woldu, T.; Raneesh, B.; Hazra, B. K.; Srinath, S.; Saravanan, P.; Reddy, M. V. R.; Kalarikkal, N. A Comparative Study on Structural, Dielectric and Multiferroic Properties of CaFe2O4/BaTiO3 Core-Shell and Mixed Composites. *J. Alloys and Compd.* **2017**, *691*, 644–652. DOI: 10.1016/j.jallcom.2016.08.277.

72. Woldu, T.; Raneesh, B.; Reddy, M. V. R.; Kalarikkal, N. Grain Size Dependent Magnetoelectric Coupling of $BaTiO_3$ Nanoparticles. *RSC Adv.* **2016**, *6*(10), 7886–7892. DOI: 10.1039/C5RA18018J.

73. Yang, L. The Effect of Magnetic Field-Tuned Resonance on the Capacitance of Laminate Composites. *J. Sens. Technol.* **2011**, *1*(03), 81–85. DOI: 10.4236/jst.2011.13011.

74. Yang, Y. T.; Li, J.; Peng, X. L.; Hong, B.; Wang, X. Q.; Ge, H. L.; Ge, H. L. Surface-Effect Enhanced Magneto-Electric Coupling in FePt/PMN-PT Multiferroic Heterostructures. *AIP Adv.* **2017**, *55833*, 0–7. DOI: 10.1063/1.4978588.

75. Yao, Z. N.; Yang, H. F.; Li, J. J.; Jiang, Q. Q.; Li, W. X.; Gu, C. Z. Detection of Domain Wall Distribution and Nucleation in Ferromagnetic Nanocontact Structures by Magnetic Force Microscopy. *J. Magn. Magn. Mater.* **2013**, *342*, 1–3. DOI: 10.1016/j.jmmm.2013.04.042.

76. Zhai, J.; Cai, N.; Shi, Z.; Lin, Y.; Nan, C.-W. Magnetic-Dielectric Properties of NiFe 2 O 4/PZT Particulate Composites. *J. Phys. D Appl. Phys.* **2004**, *37*(6), 823–827. DOI: 10.1088/0022-3727/37/6/002.

77. Zhang, H.-W.; Li, J.; Su, H.; Zhou, T.-C.; Long, Y.; Zheng, Z.-L. Development and Application of Ferrite Materials for Low Temperature Co-Fired Ceramic Technology. *Chin. Phys. B* **2013**, *22*(11), 117504. DOI: 10.1088/1674–1056/22/11/117504.

78. Zhang, R.; Jin, L.; Wu, G.; Zhang, N. Magnetic Force Driven Magnetoelectric Effect in Mn-Zn-Ferrite/PZT Composites. *Appl. Phys. Lett.* **2017**, *110*(11), 112901. DOI: 10.1063/1.4978518.

79. Zhou, J. P.; Zhang, Y. X.; Liu, Q.; Liu, P. Magnetoelectric Effects on Ferromagnetic and Ferroelectric Phase Transitions in Multiferroic Materials. *Acta Mater.* **2014**, *76*, 355–370. DOI: 10.1016/j.actamat.2014.05.038.

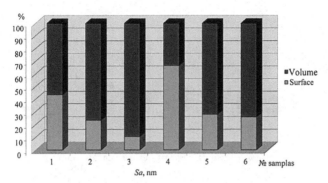

FIGURE 2.14 Height of the outer and inner surfaces of the foam box.

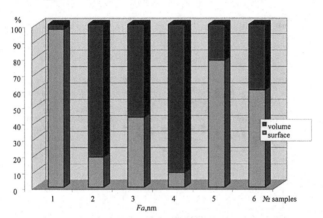

FIGURE 2.15 Strength of the probe's adhesion to the outer and inner surfaces of the foam box.

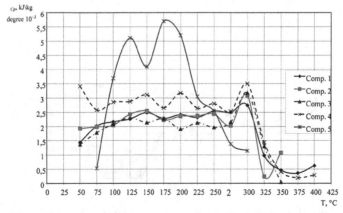

FIGURE 2.35 Dependence of heat capacity on temperature. Comp. nos. 1–5.

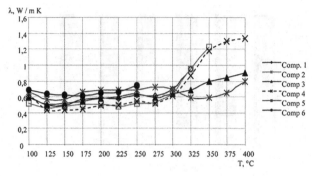

FIGURE 2.36   Dependence of thermal conductivity on temperature. Comp. nos. 1–6.

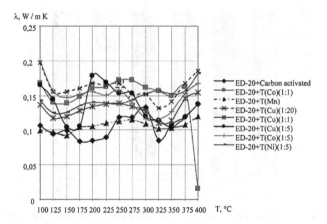

FIGURE 2.37   Dependence of thermal conductivity on temperature.

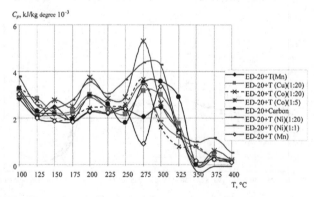

FIGURE 2.38   Dependence of heat capacity on temperature.

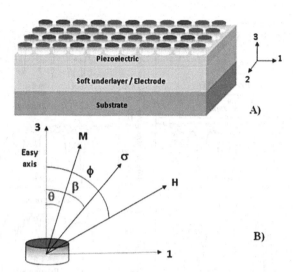

FIGURE 6.1 (a) Schematic of a patterned magnetic recording medium in which magnetic bits form a strain-mediated multiferroic system with the p iezoelectric substrate; (b) Uniaxial single domain magnetic particle in applied H field and σ(r) stress induced by the piezo-substrate when electrically activated.

Source: Reprinted from ref 70 (with permission from AIP Publishing).

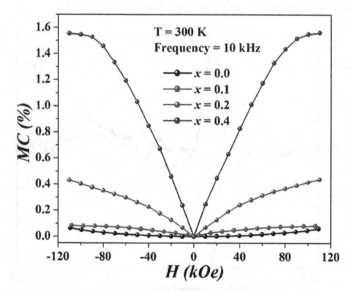

FIGURE 6.5 Magnetic field-induced change in dielectric constant at 10 kHz of the pristine BFO and NCs (x Hz of the pristine BFO and NCs t 10ed .

Source: Reprinted from ref 12 (with permission from AIP Publishing).

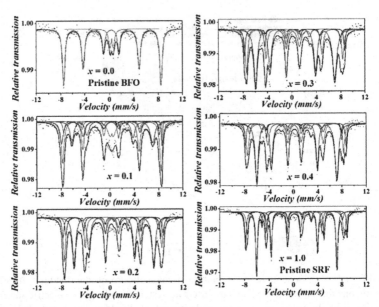

FIGURE 6.6 Room temperature Mössbauer spectra of all samples (open circles: experimental, black: fitted, blue: 4f2, green: 2a and 4f1, red: 12 k, purple: 2b, magenta: BFO, dark yellow: $Bi_2Fe_4O_9$).

Source: Reprinted from ref 12 (with permission from AIP Publishing).

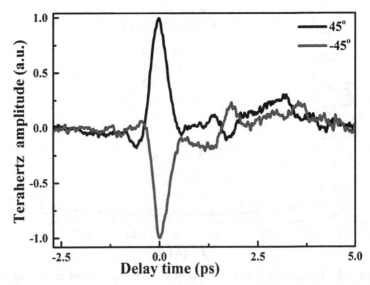

FIGURE 7.1 Semi-insulating gallium arsenide (SI-GaAs): Measured temporal profiles of THz radiation with respect to ±45° rotation angle of semiconductor surface.

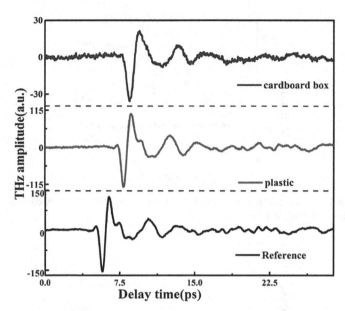

FIGURE 7.8   Transmitted THz time-domain response of reference (air), plastic lid, and cardboard box.

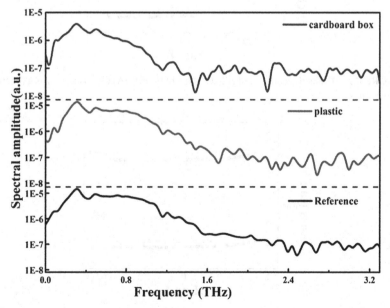

FIGURE 7.9   Transmitted frequency domain responses of reference (air), plastic lid, and cardboard box.

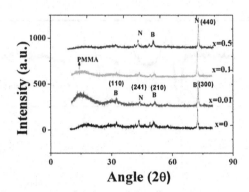

FIGURE 8.1   X-ray diffractogram of $(1-x)$ $BiFeO_3$–$xNaNbO_3$–PMMA composite film samples for $x = 0, 0.05, 0.1, 0.5$.

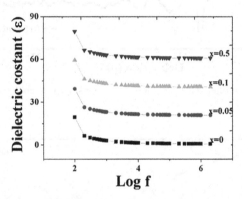

FIGURE 8.7   Dielectric constant of $BiFeO_3$–$NaNbO_3$–PMMA composite film as a function of frequency.

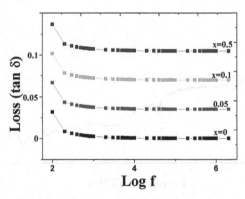

FIGURE 8.8   Dielectric loss of $BiFeO_3$–$NaNbO_3$–PMMA composite film as a function of frequency.

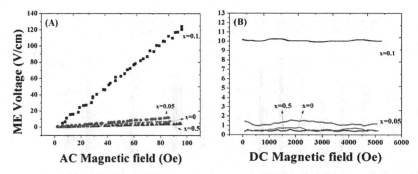

FIGURE 8.12 (a) ME voltage measured as a function of AC magnetic field, (b) ME voltage measured as a function of DC magnetic field.

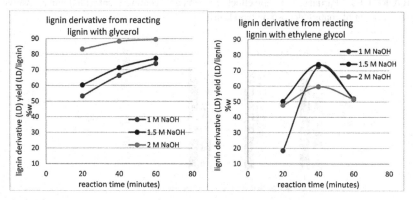

FIGURE 10.3 Relation between lignin derivative yield to the reaction time with glycerol and ethylene glycol.

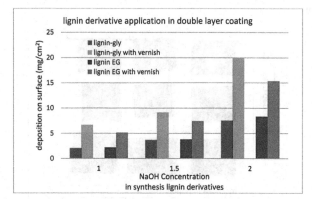

FIGURE 10.4 The effect of NaOH concentration as catalyst in the synthesis to the amount of lignin derivative deposition on the wood surface when they were applied in double-layer coating.

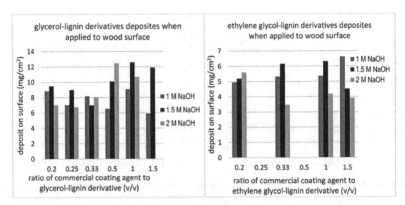

FIGURE 10.5 Partial substitution effect of commercial coating agent with lignin derivatives to the amount of coating deposit on wood surface.

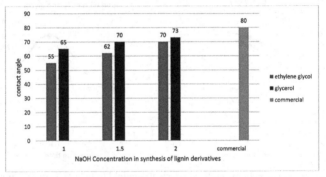

FIGURE 10.7 The effect of NaOH concentration as catalyst in the synthesis to water contact angle when they were applied in double-layer coating, compared to commercial coating agent.

FIGURE 10.8 Colors of the wood surface when glycerol–lignin derivative was applied. Increasing deposition and concentration will give darker color.

CHAPTER 7

THz GENERATION FROM GALLIUM ARSENIDE SEMICONDUCTOR SURFACES AND ITS APPLICATION IN SPECTROSCOPY

M. VENKATESH, GANESH DAMARLA, and A. K. CHAUDHARY*

Advanced Centre of Research in High Energy Materials, University of Hyderabad, Telangana 500046, India

*Corresponding author. E-mail: anilphys@yahoo.com; akcphys@gmail.com

ABSTRACT

This chapter reports the new approach for THz enhancement from semi-insulating (SI) and low-temperature gallium arsenide (LT-GaAs) semiconductor surfaces. The generated THz radiation is detected using photoconductive antennas (PCA). The obtained results from both types of semiconductor surfaces are compared and observed that the THz peak amplitude from semi-insulating gallium arsenide (SI-GaAs) was higher than the LT-GaAs surface. The effect of laser wavelength on generated THz peak amplitude is carried out. We have enhanced the efficiency of generated THz radiation from SI-GaAs using the combination of lens, BBO, and dual-wave plate (DWP). Further, the obtained THz radiation was used to record the transmittance of packing materials such as cardboard and plastic.

7.1 INTRODUCTION

The illumination of the ultrafast laser pulses on the semiconductor surface is one of the well-known optoelectronic technique used for THz emission.

Ultrafast optical excitation of semiconductors for THz generation is governed by two methods, that is, the emission from photoconductive antennas (PCA) and bare semiconductor surfaces (BSS)[1-6]. PCA devices require specific design for fabrication, external biasing arrangement, and focusing and collimating lens arrangement. Moreover, the semiconductor substrate on which antenna is supposed to be designed must have specific properties such as high resistivity, sub-picosecond carrier lifetime, high mobility, and so forth. The antenna device (small gap) cannot withstand at higher incident laser powers and biasing voltages. Additionally, it is sensitive to electrostatic charge, which causes damage to the device. On the other hand, THz pulses generated from BSS do not have such stringent requirements as mentioned for PCA. It offers freedom to select an entire surface area of BSS for the generation of THz radiation. Moreover, it can withstand even higher laser powers as compared to antennas and has simplicity of operation for THz generation. However, the power of emitted THz from semiconductors is relatively low as compared to PCA and nonlinear crystals. Hence, researchers attempted to enhance the THz radiation from BSS. Here, we attempted to enhance THz radiation emitted from semiconductors by changing experimental conditions.

Zhang et al., first demonstrated the THz generation from InP semiconductor surface in 1990[7]. Thereafter, a few groups have also explained the THz emission mechanism from semiconductor surfaces. The basic physical mechanisms involved in THz generation from semiconductor surfaces can be divided into two broad categories, that is, nonlinear and surge current effects. The bulk optical rectification and electric-field-induced optical rectification are attributed to second- and third-order nonlinear effects, while surface depletion field (SDF), Photo–Dember effect (PDE) are governed by surge currents.[3,7-11] Zhang emphasized that the SDF mechanism is responsible for THz emission, while Chaung et al. suggested that THz generation is governed by optical rectification process at higher laser intensities.[8] THz generation from narrow-gap semiconductors such as InAs, InSb, and so forth, are governed by PDE. Nakajima et al. reported that all these mechanisms can coexist in semiconductors but one of them will dominate and will be primarily responsible for THz generation. Moreover, the governing mechanism can be altered by changing the laser and substrate parameters.[9,12] In addition to these mechanisms, coherent phonon excitation, shift, and ballistic photocurrents are also responsible for THz emission.[13-16] The strength of generated THz radiation depends on several properties of semiconductors such as optical absorption, band gap, and electron and hole mobility. The important properties of semi-conductors are comprised in Table 7.1. In this chapter, we have selected semi-insulating gallium arsenide (SI-GaAs) and low-temperature gallium arsenide (LT-GaAs) semiconductor surfaces for THz generation.

TABLE 7.1 Properties of Semiconductors.

Material	Bandgap (eV)	Absorption depth* (nm)	Electron mobility (cm²/(V · s))	Hole mobility (cm²/(V · s))
GaAs	1.43	~749	8600	400
InAs	0.36	~142	30,000	240
InP	1.34	~305	4000	650
InSb	0.17	~94	76,000	3000

*at 800 nm wavelength.

Source: Adapted from ref 4.

7.2   THz EMISSION AND DETECTION

The commercially available SI-GaAs and LT-GaAs semiconductors are used as sources of THz radiation. The thickness and orientation of these semiconductors are ~0.65 mm and <110>, respectively. The emitted THz radiation is detected by dipole antenna (gap ~5 μm, length ~20 μm) using photoconductive sampling technique. The Chameleon Ultra-II oscillator (~140 fs at 80 MHz) laser system was employed in experiments. The experimental schematic used for THz generation and detection is same as mentioned by Bieler et al.[17]

Figure 7.1 depicts the emitted THz waveforms from SI-GaAs wafer at an orientation of ±45° with respect to the laser beam direction. The polarity of the emitted waveforms was changed by 180° with equal ampli-tude. The change in the polarity confirms that the direction of transient current direction is normal to the semiconductor surface.[3] The polarity of emitted THz radiation from semiconductors is governed by equation.[7]

$$E_{THz} \propto \pm \sin(\theta_r) \qquad (7.1)$$

where the incident optical angle (θ_{op}), refracted (θ_r) and transmitted angles (θ_t) of THz field through semiconductor surface follow the Fresnel law:

$$n_1(\omega_{opt})\sin(\theta_{opt}) = n_1(\omega_{ell})\sin(\theta_r) = n_2(\omega_{el2})\sin(\theta_t) \qquad (7.2)$$

By substituting the value of sin (θ_r) from Equation 7.2 to Equation 7.1, the E_{THz} becomes

$$E_{THz} \propto \pm \sin(\theta_{opt}) \qquad (7.3)$$

where $n_2(\omega_{el2})$ is the index of refraction of THz field of semiconductor material, $n_1(\omega_{opt})$ and $n_2(\omega_{el2})$ are the indices of refraction of laser beam and THz radiation outward direction, respectively. The off-normal excitation of semiconductor emits THz radiation in outward (quasi-reflection) and inward (transmitted) direction. The "+" and "−" signs in equation indicate the direction of quasi-reflected and -transmitted THz fields, respectively. In addition, the polarity of the emitted THz field depends on the sign of the angle between incident laser direction and semiconductor. The polarity of the emitted THz waveforms can also flip by illuminating the semiconductor surface with laser pulses having photon energy higher or lower than the bandgap of the material.[18,19] In addition, polarity of the THz waveform depends on incident laser intensities and the type of doping in the semiconductor material, which is governed by Photo–Dember and SDF mechanism, respectively.[3,20]

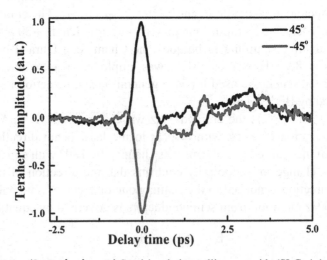

FIGURE 7.1 **(See color insert.)** Semi-insulating gallium arsenide (SI-GaAs): Measured temporal profiles of THz radiation with respect to ±45° rotation angle of semiconductor surface.

The THz peak amplitude versus laser power for SI-GaAs wafer is illustrated in Figure 7.2. The maximum amplitude obtained at 1 W laser power is nine times greater than THz amplitude at 0.1 W. The equation of THz electric field emitted from semiconductor is[7]

$$E_{THz} \propto J \propto W_{opt}[1 - R(opt)] \int E_d(x)e^{-\alpha x}dx \qquad (7.4)$$

where W_{opt} is the optical intensity of laser beam, $R(opt)$ optical reflectivity of the sample, $E_d(x)$ is static electric field, and α is the optical absorption depth. It is clear from Equation 7.4 that the amplitude of THz radiation is directly proportional to the photocurrent, which in turn depends on optical intensity of laser. However, intense laser beam cause damage to semiconductor surfaces which affect the strength of THz radiation. Therefore, the laser intensity must be kept below the damage threshold of semiconductor surface.

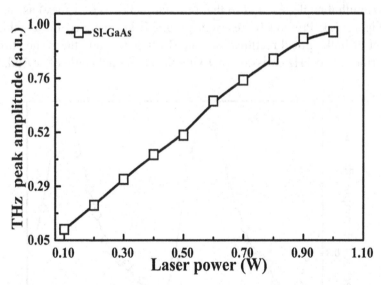

FIGURE 7.2 SI-GaAs: Measured THz amplitude with respect to incident laser power.

Figure 7.3 depicts the dependence of THz peak amplitude as a function of azimuthal rotation of <110>-oriented SI-GaAs surface. Here, the semiconductor rotation angle was fixed at 45° with respect to the laser beam direction. The azimuthal dependence of THz signal has twofold rotational symmetry (i.e., two minima and maxima). The change in THz peak amplitude with respect to azimuthal rotation of GaAs confirms the process of optical rectification for THz emission. The bulk- and surface-field-induced optical rectification process induces three- and twofold rotation symmetries, respectively, in <110>-oriented GaAs.[10,21] Present experiment also indicates the surface electric-field-induced optical rectification (SEFIOR) process is responsible for observed change in THz field amplitude with respect to rotation. For optical rectification process, the minimum value of THz amplitude must be zero for the azimuthal rotation. However, the figure shows that the minimum value of THz amplitude is 0.74 n.u., which indicates that the emitted radiation is a combination of transient current (SDF and PDE) and optical rectification mechanisms.[21,22] In broadband-gap semiconductors, SDF process is mainly responsible for THz emission rather than PDE.[20] Therefore, the combination of SEFIOR and SDF is responsible for THz emission from SI-GaAs. The SEFIOR contribution to THz emission from semiconductors can be minimized by proper selection of azimuthal angle referred to the minimum value (i.e., valley) as shown in Figure 7.3. Previously, research groups have removed the combined effect of bulk optical rectification and SEFIOR to study the surge current contribution to THz emission by fixing the azimuthal angle of crystal.[22-24]

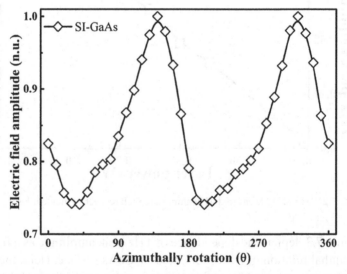

FIGURE 7.3 SI-GaAs: Dependence of THz field amplitude with respect to emitter azimuthal rotation.

7.2.1 COMPARATIVE STUDIES OF SEMI-INSULATING (SI) AND LOW-TEMPERATURE GALLIUM ARSENIDE (LT-GAAS)

Figure 7.4 shows the THz peak amplitude from SI and LT-GaAs surfaces as a function of incident laser power. The measured value of amplitude is 77.1 a.u., 65.2 a.u. for SI and LT-GaAs semiconductors at 1000 mW. Therefore, the THz peak amplitude from SI-GaAs is 1.2 times greater than LT-GaAs semiconductor. THz peak amplitude from SI and LT-GaAs semiconductors at 100 mW power is 8.52 and 5.78 a.u., respectively. Thus, the measured THz peak amplitude from SI and LT-GaAs wafers at 1 W was 9.04 and 11.28 times higher than generated amplitude at 100 mW laser power. Moreover, the power of generated THz radiation from SI-GaAs was more than LT-GaAs wafer. In principle, the SDF mechanism is responsible for THz generation in large bandgap semiconductors (e.g., GaAs, InP).[20] The radiated THz electric field (E_{THz}) due to SDF mechanism is given by following equation.[25]

$$E_{THz} = \frac{dJ}{dt} \propto \mu_e E_{static} \propto \mu_e \qquad (7.5)$$

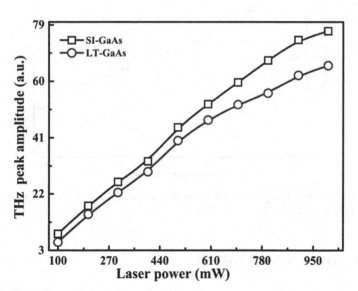

FIGURE 7.4 Dependence of THz peak amplitude from SI and LT-GaAs as function of incident laser power.

Where J is current density, μ_e is mobility of electrons, and E_{static} represents the depletion field present in semiconductors. It is clear from Equation 7.5 that the THz emission from semiconductors depends on carrier mobility and inbuilt electric field. Since the electron mobility of SI-GaAs is higher than the LT-GaAs wafer, it is one of the main reasons for higher order THz electric field from SI-GaAs.[26]

The dependence of THz peak amplitude as a function of incident laser wavelength is depicted in Figure 7.5. The amplitude of generated THz radiation is decreasing with an increase in the incident laser wavelength between 770 and 830 nm. Previously, Zhang et al. reported the decrement of obtained THz peak amplitude from GaAs surfaces by decreasing the incident laser photon energy (longer wavelengths) with off-normal and normal illumination.[18,19] Arlauskas et al. explained that the enhancement in carrier separation is responsible for the rise in THz peak amplitude with an increase in laser photon energy.[27]

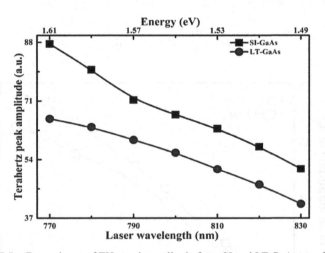

FIGURE 7.5 Dependence of THz peak amplitude from SI and LT-GaAs as a function of laser central wavelength.

7.2.2 THz ENHANCEMENT

Previously, several groups have attempted to enhance the efficiency of generated THz radiation generated from semiconductor surfaces by changing the laser parameters, applying magnetic field, and by modifying the substrate properties.[16,23,28–35] Cote et al. generated broadband THz radiation from GaAs surfaces using the combination (quantum interference) of 1550 and 775 nm wavelengths.[16] Similarly, other groups have also reported the enhancement in plasma-based THz generation using similar experimental setup reported by Cote et al.[38,39] Here, we achieved enhanced THz emission SI-GaAs surface by incorporating type-I second harmonic cut BBO crystal and a dual-wave plate (DWP). The experimental schematic used for enhancing THz emission from semiconductors and detection by pyroelectric detector is illustrated in Figure 7.6. The experimentally measured noise equivalent power of pyroelectric detector is ~2 nW at room temperature. We have used three different configurations, that is, focusing lens, lens and BBO combination, and lens and BBO combined with DWP for enhancement in THz radiation. Figure 7.6 depicts the schematic of three different configurations used for generation and enhancement of THz signal. The BBO crystal has thickness ~0.2 mm and phase matching angle ~29.2°. The focal length of the employed plano-convex lens is 30 cm. The DWP works as a half-wave plate ($\lambda/2$) and wave plate (λ) for 800 and 400 nm laser wavelengths, respectively. This implies that the polarization of 800 and 400 nm undergoes 90 and 0° rotation when it is rotated to 45° with respect to input beam polarization. The BBO has reflection losses of the order 18% at 800 nm incident laser wavelength. The constant input laser power is maintained during the course of entire experiment and power ratio of 800:400 nm (from BBO crystal) is about ~60:40.

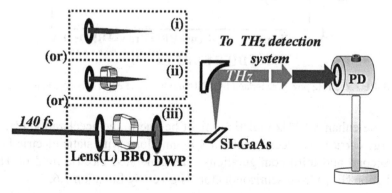

FIGURE 7.6 Schematic of three experimental configurations used for THz emission from SI-GaAs wafer (where i: using focusing lens, ii: optically aligned lens and BBO crystal, and iii: optically aligned lens, BBO, and dual-wave plate [DWP] arrangements, PD: pyroelectric detector).

Three experimental arrangements are shown in Figure 7.6 and are marked as first, second, and third configurations. The generated THz power in case of second and third configurations is maximized by means of proper orientation of BBO and DWP. The strength of generated THz radiation from semiconductors in the abovementioned configurations with oscillator laser pulses at 800 nm incident laser wavelength is illustrated in Figure 7.7. The recorded THz powers in first, second, and third configurations are 13.3, 14.7, and 17.1 nW, respectively. The radiation obtained from SI-GaAs wafer in third configuration was ~1.285 and ~1.16 times higher than the power obtained in first and second configurations. Our observation revealed that the third configuration demonstrated the highest THz radiation as compared to second and first configurations (lens, BBO, and DWP [third] > lens and BBO combination [second] > lens [first]).

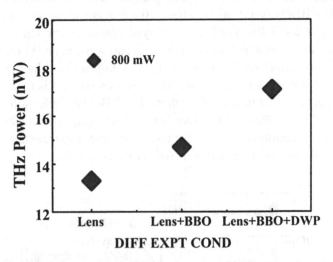

FIGURE 7.7 THz power generated from SI-GaAs in three experimental conditions.

The enhanced THz radiation might be due to the acceleration of free charge carriers by electric field of 400 nm along with static electric field in second and third configurations. The generated electric field of THz radiation (E_{THz}) from semiconductors is given by Equation 7.6.[33]

$$E_{THz} = \frac{dJ}{dt} \propto \frac{d(ne\vartheta)}{dt} \propto \frac{d\vartheta}{dt} \propto a \propto neE \propto E \tag{7.6}$$

where J is the current density, n is charge density, e is the electronic charge, ϑ is the velocity of electrons, E is the total electric field, and a (or) d ϑ/dt represents the acceleration of electrons under the influence of electric field. From Equation 7.6 the radiated electric field of THz radiation depends on the carrier acceleration, which is ultimately governed by applied/inbuilt electric field of semiconductor. Therefore, THz enhancement is due to acceleration of carriers (created by 800 nm) by 400 nm electric field along with static field in case of second configuration; it is further enhanced due to the enhanced acceleration of carriers through the parallel alignment of 800 and 400 nm polarizations by DWP in third configuration.

7.2.3 THz SPECTROSCOPY

THz spectrometer was built using semiconductor (generator), and PCA (detector) was used to measure the THz transmittance of cardboard box of length 12 cm filled with plastic gloves and plastic lid of thickness 1.5 cm. The transmitted THz time-domain response of reference, plastic lid, and cardboard box is shown in Figure 7.8. The transmitted THz peak–peak amplitudes of reference, plastic, and rubber were 176 a.u., 137 a.u., and 59 a.u., respectively. The fast Fourier transform of THz temporal profiles is shown in Figure 7.9 and the emitted frequency spectrum was extended up to 2.5 THz.

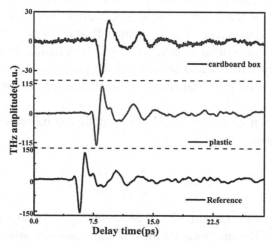

FIGURE 7.8 (See color insert.) Transmitted THz time-domain response of reference (air), plastic lid, and cardboard box.

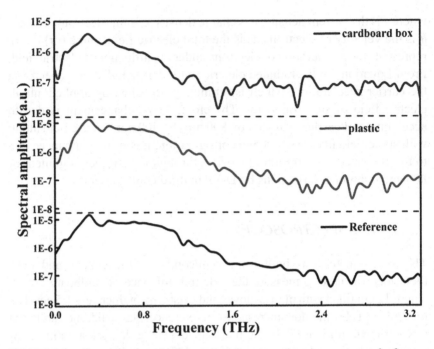

FIGURE 7.9   (See color insert.) Transmitted frequency domain responses of reference (air), plastic lid, and cardboard box.

The Fourier transform of THz temporal profile provides information on properties of the samples such as refractive index, absorption, and transmission coefficient without using Kramer-Kronig relations for analysis. The transmitted electric field of THz radiation through samples was altered by intrinsic and absorption of samples under consideration. Here, we measured the transmission coefficient of cardboard box and plastic lid.

Transmission coefficient of THz amplitude can be retrieved from the following equation.[40,41]

$$T(\nu)_{THz} = \frac{E_{THz\text{-}Sam}(\nu)}{E_{THz\text{-}Ref}(\nu)} \tag{7.7}$$

where $T(\nu)_{THz}$ is the transmission coefficient, $E_{THz\text{-}Sam}(\nu)$ and $E_{THz\text{-}Ref}(\nu)$ are the obtained THz spectrum when THz radiation is transmitted through sample and reference.

Figure 7.10 depicts the transmittance of THz radiation through cardboard box and plastic lid. The transmittance of plastic and cardboard box decreases with respect to an increase in THz frequency. The THz transmittance of cardboard box and plastic was higher at 0.17 THz (i.e., 0.498 a.u.) and 0.32 THz (i.e., 0.7362 a.u.), respectively. The transmittance of cardboard box was reduced to zero at 1.5 THz as shown in Figure 7.10.

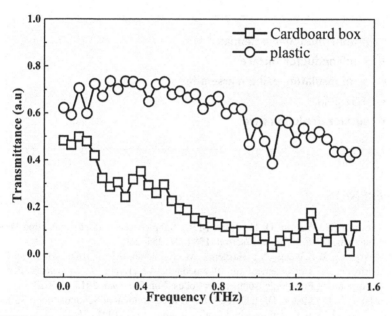

FIGURE 7.10 THz transmission coefficients of plastic lid and cardboard box.

7.3 CONCLUSION

The THz radiation was successfully generated and enhanced from semiconductor surfaces by varying the experimental configurations. The enhanced THz radiation was higher in case of third configuration (BBO + DWP + lens) as compared to second (BBO + DWP) and first (lens) configuration. THz peak amplitude is higher in SI-GaAs as compared to LT-GaAs due to its high mobility. The effect of laser wavelength on GaAs semiconductors reveals that THz peak amplitude is decreasing with an increase in laser wavelengths. In addition, we successfully recorded the THz transmittance of plastic and cardboard box.

ACKNOWLEDGMENT

We gratefully acknowledge the financial support (under Phase-III grants) given by the DRDO, Ministry of Defence, Government of India, India.

KEYWORDS

- photoconductive antennas
- semiconductor surface
- semi-insulating gallium arsenide
- THz peak
- surface depletion field

REFERENCES

1. Smith, P. R.; Auston, D. H.; Nuss, M. C. Subpicosecond Photoconducting Dipole Antennas. *IEEE J. Quant. Electron.* **1988**, *24*, 255–260.
2. Berry, C. W.; Wang, N.; Hashemi, M. R.; Unlu, M.; Jarrahi, M. Significant Performance Enhancement in Photoconductive Terahertz Optoelectronics by Incorporating Plasmonic Contact Electrodes. *Nat. Commun.* **2013**, *4*, 1622.
3. Zhang, X. C.; Auston, D. H. Optoelectronic Measurement of Semiconductor Surfaces and Interfaces with Femtosecond Optics. *J. Appl. Phys.* **1992**, *71*, 326–338.
4. Sakai, K.; Kikō, J. T. K. *Terahertz optoelectronics.* Springer: Berlin, 2005.
5. Venkatesh, M.; Rao, K. S.; Abhilash, T. S.; Tewari, S. P.; Chaudhary, A. K. Optical Characterization of GaAs Photoconductive Antennas for Efficient Generation and Detection of Terahertz Radiation. *Opt. Mater.* **2014**, *36*, 596–601.
6. Mottamchetty, V.; Chaudhary, A. K. Improvised Design of THz Spectrophotometer Using LT-GaAs Photoconductive Antennas, Pyroelectric Detector and Band-Pass Filters. *Indian J. Phys.* **2016**, *90*, 73–78.
7. Venkatesh, M.; Chaudhary, A. K. In Generation of THz radiation from Low temperature Gallium Arsenide (LT-GaAs) photoconductive (PC) antennas using tunable femtosecond oscillator, *12th International Conference on Fiber Optics and Photonics*, Optical Society of America: Kharagpur, 2014/12/13 2014; p S5A.33.
8. Venkatesh, M.; Chaudhary, A. K.; Rao, K. S. In *Generation and detection of Terahertz radiation from low temperature gallium arsenide photoconductive antennas using autocorrelation technique.* Recent Advances in Photonics (WRAP), 2013 Workshop on, Dec 17–18, 2013; pp 1–2.

9. Zhang, X. C.; Hu, B. B.; Darrow, J. T.; Auston, D. H. Generation of Femtosecond Electromagnetic Pulses from Semiconductor Surfaces. *Appl. Phys. Lett.* **1990**, *56*, 1011–1013.

10. Chuang, S. L.; Schmitt-Rink, S.; Greene, B. I.; Saeta, P. N.; Levi, A. F. Optical Rectification at Semiconductor Surfaces. *Phys. Rev. Lett.* **1992**, *68*, 102.

11. Nakajima, M.; Oda, Y.; Suemoto, T. Competing Terahertz Radiation Mechanisms in Semi-Insulating InPat High-Density Excitation. *Appl. Phys. Lett.* **2004**, *85*, 2694–2696.

12. Radhanpura, K.; Hargreaves, S.; Lewis, R. A. Bulk and Surface Field-Induced Optical Rectification from (11N) Zincblende Crystals in a Quasireflection Geometry. *Phys. Rev. B* **2011**, *83*, 125322.

13. Krotkus, A.; Bertulis, K.; Kai, L.; Xu, J.; Zhang, X. C. Terahertz Emission from the Structures Containing Low-Temperature-Grown GaAs Layers. *Semicond. Sci. Technol.* **2004**, *19*, S452.

14. Heyman, J. N.; Coates, N.; Reinhardt, A.; Strasser, G. Diffusion and Drift in Terahertz Emission at GaAs Surfaces. *Appl. Phys. Lett.* **2003**, *83*, 5476–5478.

15. Tani, M.; Fukasawa, R.; Abe, H.; Matsuura, S.; Sakai, K.; Nakashima, S. Terahertz Radiation from Coherent Phonons Excited in Semiconductors. *J. Appl. Phys.* **1998**, *83*, 2473–2477.

16. Côté, D.; Laman, N.; van Driel, H. M. Rectification and Shift Currents in GaAs. *Appl. Phys. Lett.* **2002**, *80*, 905–907.

17. Bieler, M.; Pierz, K.; Siegner, U. Simultaneous Generation of Shift and Injection Currents in (110)-grown GaAs/AlGaAs Quantum Wells. *J. Appl. Phys.* **2006**, *100*, 083710.

18. Côté, D.; Fraser, J. M.; DeCamp, M.; Bucksbaum, P. H.; van Driel, H. M. THz Emission from Coherently Controlled Photocurrents in GaAs. *Appl. Phys. Lett.* **1999**, *75*, 3959–3961.

19. Venkatesh, M.; Ramakanth, S.; Chaudhary, A. K.; Raju, K. C. J. Study of Terahertz Emission from Nickel (Ni) Films of Different Thicknesses Using Ultrafast Laser Pulses. *Opt. Mater. Express* **2016**, *6*, 2342–2350.

20. Hu, B.; Zhang, X.-C.; Auston, D. Terahertz Radiation Induced by Subband-Gap Femtosecond Optical Excitation of GaAs. *Phys. Rev. Lett.* **1991**, *67*, 2709.

21. Zhang, X.-C.; Jin, Y.; Yang, K.; Schowalter, L. Resonant Nonlinear Susceptibility Near the GaAs Band Gap. *Phys. Rev. Lett.* **1992**, *69*, 2303.

22. Apostolopoulos, V.; Barnes, M. E. THz Emitters Based on the Photo-Dember Effect. *J. Phys. D: Appl. Phys.* **2014**, *47*, 374002.

23. Radhanpura, K. All-Optical Terahertz Generation from Semiconductors: Materials and Mechanisms. Doctor of Philosophy Thesis, University of Wollongong, 2012.

24. Gu, P.; Tani, M.; Kono, S.; Sakai, K.; Zhang, X.-C. Study of Terahertz Radiation from In As and InSb. *J. Appl. Phys.* **2002**, *91*, 5533–5537.

25. Wu, X.; Xu, X.; Lu, X.; Wang, L. Terahertz Emission from Semi-Insulating GaAs with Octadecanthiol-Passivated Surface. *Appl. Surf. Sci.* **2013**, *279*, 92–96.

26. Inoue, R.; Takayama, K.; Tonouchi, M. Angular Dependence of Terahertz Emission from Semiconductor Surfaces Photoexcited by Femtosecond Optical Pulses. *J. Opt. Soc. Am. B* **2009**, *26*, A14–A22.

27. Hwang, J.-S.; Lin, H.-C.; Chang, C.-K.; Wang, T.-S.; Chang, L.-S.; Chyi, J.-I.; Liu, W.-S.; Chen, S.-H.; Lin, H.-H.; Liu, P.-W. The Dependence of Terahertz Radiation on the Built-in Electric Field in Semiconductor Microstructures. *Opt. Express* **2007**, *15*, 5120–5125.

28. Tani, M.; Matsuura, S.; Sakai, K.; Nakashima, S.-I. Emission Characteristics of Photoconductive Antennas Based on Low-Temperature-Grown GaAs and Semi-Insulating GaAs. *Appl. Opt.* **1997**, *36*, 7853–7859.

29. Andrius, A.; Arūnas, K. THz Excitation Spectra of AIIIBV Semiconductors. *Semicond. Sci. Technol.* **2012**, *27*, 115015.

30. McBryde, D.; Barnes, M.; Berry, S.; Gow, P.; Beere, H.; Ritchie, D.; Apostolopoulos, V. Fluence and Polarisation Dependence of GaAs Based Lateral Photo-Dember Terahertz Emitters. *Opt. Express* **2014**, *22*, 3234–3243.

31. Antsygin, V. D.; Mamrashev, A. A.; Nikolaev, N. A.; Potaturkin, O. I. Effect of a Magnetic Field on Wideband Terahertz Generation on the Surface of Semiconductors. *IEEE Trans. Terahertz Sci. Technol.* **2015**, *5*, 673–679.

32. Kang, C.; Woo Leem, J.; Wook Lee, J.; Su Yu, J.; Kee, C.-S. Characteristics of Terahertz Pulses from Antireflective GaAs Surfaces with Nanopillars. *J. Appl. Phys.* **2013**, *113*, 203102.

33. Prieto, E. A. P.; Vizcara, S. A. B.; Somintac, A. S.; Salvador, A. A.; Estacio, E. S.; Que, C. T.; Yamamoto, K.; Tani, M. Terahertz Emission Enhancement in Low-Temperature-Grown GaAs with an n-GaAs Buffer in Reflection and Transmission Excitation Geometries. *J. Opt. Soc. Am. B* **2014**, *31*, 291–295.

34. Atrashchenko, A.; Arlauskas, A.; Adomavičius, R.; Korotchenkov, A.; Ulin, V. P.; Belov, P.; Krotkus, A.; Evtikhiev, V. P. Giant Enhancement of Terahertz Emission from Nanoporous GaP. *Appl. Phys. Lett.* **2014**, *105*, 191905.

35. Shi, Y.; Xu, X.; Yang, Y.; Yan, W.; Ma, S.; Wang, L. Anomalous Enhancement of Terahertz Radiation from Semi-Insulating GaAs Surfaces Induced by Optical Pump. *Appl. Phys. Lett.* **2006**, *89*, 81129 (1–3).

36. Jaculbia, R.; Balgos, M.; Mangila, N.; Tumanguil, M.; Estacio, E.; Salvador, A.; Somintac, A. Enhanced Terahertz Emission from GaAs Substrates Deposited with Aluminum Nitride Films Caused by High Interface Electric Fields. *Appl. Surf. Sci.* **2014**, *303*, 241–244.

37. Wu, X.; Quan, B.; Xu, X.; Hu, F.; Lu, X.; Gu, C.; Wang, L. Effect of Inhomogeneity and Plasmons on Terahertz Radiation from GaAs (1 0 0) Surface Coated with Rough Au Film. *Appl. Surf. Sci.* **2013**, *285*, 853–857.

38. Minami, Y.; Kurihara, T.; Yamaguchi, K.; Nakajima, M.; Suemoto, T. High-Power THz Wave Generation in Plasma Induced by Polarization Adjusted Two-Color Laser Pulses. *Appl. Phys. Lett.* **2013**, *102*, 41105 (1–4).

39. Dai, J.; Karpowicz, N.; Zhang, X. C. Coherent Polarization Control of Terahertz Waves Generated from Two-Color Laser-Induced Gas Plasma. *Phys. Rev. Lett.* **2009**, *103*, 023001 (1-4).

40. Chen, J.; Chen, Y.; Zhao, H.; Bastiaans, G. J.; Zhang, X.-C. Absorption Coefficients of Selected Explosives and Related Compounds in the Range of 0.1–2.8 THz. *Opt. Express* **2007**, *15*, 12060–12067.

41. Gowen, A.; O'Sullivan, C.; O'Donnell, C. Terahertz Time Domain Spectroscopy and Imaging: Emerging Techniques for Food Process Monitoring and Quality Control. *Trends Food Sci. Technol.* **2012**, *25*, 40–46.

CHAPTER 8

MULTIFERROIC POLYMER FILM COMPOSITES FOR MEMORY APPLICATION

REHANA P. UMMER[1,*] and NANDAKUMAR KALARIKKAL[1,2]

[1]School of Pure and Applied Physics, Mahatma Gandhi University, Kottayam, Kerala 68660, India

[2]International and Inter University Centre for Nanoscience and Nanotechnology, Mahatma Gandhi University, Kottayam, Kerala 686560, India

*Corresponding author. E-mail: rehana2009spap@gmail.com

ABSTRACT

Polymer host multiferroic nanocomposites are appropriate for device application because they are highly flexible and can be cast in thin film form. Ceramic polymer composites have become potential candidates for integration into high-frequency electronics. For many applications including integral thin film capacitors, electrostriction systems, for artificial muscles and electric stress control devices, apart from a high dielectric constant, a low dielectric loss is also required. In particular, one-dimensional ferroelectric materials have recently been studied extensively in the quest to miniaturize devices for nonvolatile random access memory applications and discovering interesting physical phenomena at nanoscale. Fabrication of polymer fibers with diameters in the range of micrometers and nanometers has attracted the attention of several researchers in recent times all over the globe. Significant improvement in the magnetoelectric (ME) response is observed in polymer-based nanocomposite films. Polymethyl methacrylate (PMMA) has been chosen as the matrix due to low dielectric loss of the polymer. In this chapter, the fabrication and characterization of $BiFeO_3$–$NaNbO_3$–PMMA films are detailed. The PMMA–$BiFeO_3$–$NaNbO_3$ films were prepared by solvent casting method. Structural analysis was performed using X-ray diffraction, scanning electron microscope, atomic force microscopy, and confocal Raman spectroscopy. The ME coupling studies confirmed the multiferroic nature of the film composites at room temperature. The magnetization measurements have shown a weak ferromagnetic behavior at room temperature for both film and fiber composites. The dielectric study of the composite shows low dielectric loss compared to bulk $BiFeO_3$. The electric property is due to the ferroelectric phase of $NaNbO_3$ and poly (methyl methacrylate). The present findings may lead to an easy and cheap method for preparing flexible energy storing and transforming materials which are required in embedded-capacitors, microelectromechanical systems, ultrasonic resonators, high-power transducers, actuators, and so on.

8.1　INTRODUCTION

Many new multiferroic materials with coupled antiferromagnetic and ferroelectric orders have been discovered recently. However, there are very few systems with ordering temperatures above 300 K.

Effective conversion of ubiquitous mechanical energy into electricity is one of the most important issues in the scientific community. In particular, the need for self-powering nanodevices from tiny vibrations such as air pressure and heartbeat has inspired the interest for developing high-performance piezoelectric nanomaterials.

Until now, piezoelectric ZnO nanowires have been the most outstanding candidate for nanogenerators. Among piezoelectric materials, displacive ferroelectrics such as lead zirconate titanate exhibit very large piezoelectric coefficients. However, the use of ferroelectric material for harvesting mechanical energy is rare, possibly due to the difficulty in forming one-dimensional nanostructures and the elevated temperature required for the growth.

Multiferroic tunnel junctions have become the objects of much scientific importance for their promising applications in spintronic devices for nonvolatile memory and sensor applications, especially magnetic and ferroelectric ones. Magnetic nanoparticles with functionalized surfaces have recently found numerous applications in biology, medicine, and biotechnology. Some application concern in vitro applications such as magnetic tweezers and magnetic separation of proteins and DNA molecules. Recently, researchers found out piezoelectric nanogenerators fabricated using lead-free $NaNbO_3$ nanowires. Using a piezoelectric polymer as the matrix for nanoparticle and improvement in the electrical property is expected and thus multiferroicity could be highly enhanced.

Polymer host multiferroic nanocomposites are appropriate for device application because they are highly flexible and can be cast in thin film form.[1] Several other problems such as high leakage current and dielectric loss, which lead to failure during operation can be eliminated in these polymer-based composites.[1,2] Polymethyl methacrylate (PMMA) is an important member in the family of polyacrylic and methacrylic esters. PMMA has several desirable properties including exceptional optical clarity, good weather, high strength, and excellent dimensional stability. PMMA nanocomposites offer the potential for reduced gas permeability, improved physical performance, and increased heat resistance without a sacrifice in optical clarity. Among the insulating polymers, PMMA is a very good insulating polymer. The dielectric constant and dielectric loss of this polymer have been reported to be ~2.6 and 0.014, respectively at 1 MHz frequency. The low dielectric loss of this polymer enables it to be used as a matrix in nanocomposites.

Ahlawat et al. had studied the magnetic and dielectric properties of PMMA-BFO nanocomposites. They found that with an increase of the BFO loading in the composite, the magnetic as well as the dielectric properties were enhanced.[3] Sun et al. had reported that with an increase of the BFO loading in PMMA, the microwave shielding ability of the composite was increased.[4]

8.2 SCOPE OF RESEARCH ON MULTIFERROICS

Nowadays, miniaturization of microwave devices is important for communication systems. Small size and high-performance devices are necessary to reduce the cost and improve the functionality of a system. There is a need for frequency tunable devices such as resonators, phase shifters, delay lines, and filters in the microwave and millimeter wave frequency regimes. Ferrites are used in tunable microwave and millimeter wave devices and the tunability is traditionally realized through the variation of a bias magnetic field. This magnetic tuning could be achieved over a very wide frequency range but is relatively slow and noisy and requires high power for operation. Similar devices but with some unique advantages could be realized by replacing the ferrite with a ferrite–ferroelectric composite. The structures and models provided in this issue can be used to develop components that not only reduce the cost but also improve the performance and application regimes. The analytical expressions for the magnetoelectric (ME) coefficient of magnetostrictive–piezoelectric laminates have been mostly derived under the assumption of homogeneity of electric, magnetic, elastic fields, and by employing boundary conditions for mechanical stress at the structure facets. ME materials have been investigated to find applications in sensors, transducers, actuators, energy harvesters, and servomechanism. The coexistence of ferroelectric and magnetic ordering in multiferroics (MF) has led to the development of various kinds of materials and conceptions of many novel applications such as the development of a memory device utilizing the multifunctionality of the multiferroic materials leading to a multistate memory device with electrical writing and non-destructive magnetic reading operations. In the commercially available magnetic memory device (MRAM), data is written by switching the magnetic states ($\pm M$) upon application of a magnetic field while to read data, one exploits variation of magnetoresistance in the magnetic states.[5,6]

Being hard ferromagnets, the materials in the MRAMs possess high coercivity[7] resulting in the large magnetic field and thus consume the large amount of energy. In contrast, ferroelectric random-access memories (FeRAMs) possess faster writing speeds through polarization switching and are energy efficient. However, these have limitations on their size and show slow readability due to their destructive read operation and subsequent reset.[8]

Therefore, a memory device with the best functionalities of FeRAMs and MRAMs (ferroelectric write and magnetic read operations) would effectively enhance the writing speed and reduce the specific energy consumption. MF and ME with the simultaneous presence of electrical and magnetic order parameters can make use of both the functionalities of FERAMs and MRAMs, independently to store binary data. Further, the coupling between electrical ($\pm P$) and magnetic ($\pm M$) order parameters in these materials provides the possibility of possessing additional functionalities such as electrical (magnetic) control of magnetization (polarization) enabling design of futuristic multistate memory devices with electrical writing and nondestructive magnetic reading operations.[9–11]

However, with strong coupling between the electric and magnetic states (ME coupling) the available switchable states namely (+P, +M), (+P, −M), (−P, +M), and (−P, −M) are not absolutely independent of each other and only combinations that are independently achievable are either (+P, +M) and (−P, −M) or (+P, −M), and (−P, +M).[12] Thus, the device is restricted to two states similar to conventional ferroelectric or magnetic memories.[13] However, as recent research has shown, this problem can be circumvented by forming a ferromagnetic–ME tunnel junction where the combination of electroresistance and magnetoresistance can result in four-state memory effect.[14]

8.3 BIFEO$_3$ AS MEMORY DEVICE

The most promising and most studied material for memory devices is bismuth ferrite, BFO which shows room temperature ferroelectricity (Tc ≈ 1100 K) and antiferromagnetism (TN ≈ 640 K). The material has a rhombohedral structure and shows a large ferroelectric polarization[15] with spontaneous polarization Ps vector along (111) axis and significant ME coupling.[16] Owing to its large spontaneous polarization, possibly the largest among all known perovskite and nonperovskite multiferroic oxides, BFO is also a prospective candidate for next-generation ferroelectric memory applications. However, the major challenges BFO faces in this context is its poor leakage characteristics, the tendency to fatigue[17] and thermal decomposition near coercive field.[18]

These limitations are partly circumvented by the composite materials. Further BFO has been reported to emit THz radiation when irradiated with a femtosecond laser pulse which may have huge potential in telecommunication applications.[19,20] Moreover, THz emission is dependent on the poling state of BFO, and therefore, ultrafast, nondestructive ferroelectric memory readout is possible. Additionally, high-frequency operation in THz range eliminates leakage.[21]

However, the major thrust toward the research of BFO and related materials are driven by their prospective applications in ME and spintronic devices where they are primarily used as memory elements. The key advantages of BFO-based memory devices are their electrical writing and magnetic reading operations which can further utilize the advantages of solid-state circuits such as their low energy consumption, scalability, nondestructive read operations, and so forth.[22]

8.4   MULTIFERROIC COMPOSITE FILMS

To overcome the problems with single phase MF, composite MF are being developed and it has been found that a much stronger ME coupling effect is realized in a composite of piezoelectric phase and magnetostrictive phase.[23] Multiferroic thin films are technologically important in spintronic devices due to low power consumption.[24–30] Multiferroic thin films and nanostructures have been produced using a wide variety of techniques including sputtering, spin coating, solgel process, metal organic chemical vapor deposition, molecular beam epitaxy, pulsed laser deposition, and more.[31,32] The growth defects in the films may either have a favorable or an unfavorable effect for the intended applications. Compared with bulk ME ceramics, multiferroic thin films have unique superiorities. Among single phase, thin film MF, hexagonal manganites, and bismuth-based perovskites are the most studied materials. The first investigated multiferroic thin film was the hexagonal manganite $YMnO_3$.[33]

8.5   STUDIES ON $BIFEO_3$–$NANBO_3$–PMMA COMPOSITE FILM SAMPLES

We report BFO/PMMA nanocomposite films which were prepared by introducing $BiFeO_3$–$NaNbO_3$ nanoparticles in an insulating and diamagnetic PMMA polymer matrix so that each nanoparticle in polymer can behave like an active element keeping its magnetic properties intact. In these composite films, magnetic and dielectric properties depend on the concentration of filler particles. Hence, magnetic and dielectric properties were investigated in nanocomposite films with varying volume fraction of nanoparticles.

8.5.1   PREPARATION OF THE FILM

Initially, $(1-x)$ $BiFeO_3$-$xNaNbO_3$ ceramic powder for $x = 0$, 0.05, 0.1, and 0.5 were prepared by Pechini method.[34] PMMA–$BiFeO_3$–$NaNbO_3$ films were prepared by solvent casting method. PMMA solution is prepared by using acetone as the solvent. The ceramic nanoparticles (5 wt.%) were then dispersed in PMMA solution and then ultrasonicated for 30 min. A gelatinous brownish white solution was obtained. The obtained solution is poured into Petri dishes and kept at room temperature for 2–3 days. Within this time the desired films were formed. The crystal structures of the samples were examined by Phillips X'Pert PANalytical X-ray diffraction (XRD) with Cu-Kα radiation (1.54056 Å). Step scanned powder XRD data were collected in the 2θ range 10–80° at room temperature. The detailed structural analysis was performed using scanning electron microscope (SEM) (JEOL JSM 6390), and confocal Raman spectroscopy. A conventional ME measurement has been carried out using the lock-in amplifier method and room temperature dielectric studies were performed using an Agilent E4980A precision LCR meter. The magnetization measurements were performed using vibrating sample magnetometer (VSM).

8.5.2 X-RAY DIFFRACTION STUDIES

Figure 8.1 shows X-ray diffractograms at room temperature of the PMMA–BiFeO$_3$–NaNbO$_3$ composite films. The broad peak at 13° is due to the presence of amorphous polymer chain and other peaks correspond to ceramic powder. The peaks which are marked by B and N corresponds to BiFeO$_3$ and NaNbO$_3$, respectively. The strong diffraction peak at 70° is shifted to 71° when x increases to 0.5. The peak at 70° corresponds to (300) plane of BiFeO$_3$ plane while the peak at 71° corresponds to (440) plane of NaNbO$_3$. By increasing x value, the content of BiFeO$_3$ decreases and that of NaNbO$_3$ increases. The average particle sizes were calculated using Scherrer equation and it lies between 100 and 150 nm for all compositions.

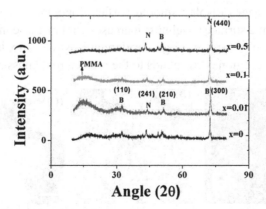

FIGURE 8.1 (See color insert.) X-ray diffractogram of (1−x) BiFeO$_3$–xNaNbO$_3$–PMMA composite film samples for x = 0, 0.05, 0.1, 0.5.

8.5.3   SCANNING ELECTRON MICROSCOPY

SEM images are shown in Figure 8.2. The black surface is the polymer matrix. No pores were observed in the film. We can see the single particles as well as the agglomerated particles in the SEM images. Eventhough nanoparticle agglomerates are there, the dispersion of the nanoparticle is fairly uniform. The grain size is found to be increasing when x increases from x=0.1 to x=0.5. Figure 8.2 suggests that grain size increases with increase in $NaNbO_3$ concentration.

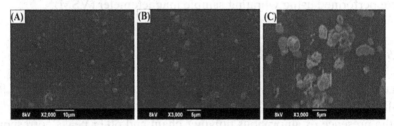

FIGURE 8.2   SEM images of (1−x) BiFeO3–xNaNbO3–PMMA composite film samples for (a) x=0, (b) x=0.1, (c) x=0.5.

8.5.4   ATOMIC FORCE MICROSCOPY

The two-dimensional and three-dimensional AFM images of $BiFeO_3$–$NaNbO_3$–PMMA composite film samples are shown in Figures 8.3 and 8.4, respectively. For x=0 composition, the two- dimensional and three-dimensional images display smooth surface topography while for other compositions the surface roughness increases with increase in x. It can also be observed that the area of bright portion increases when x increases. The bright and dark region corresponds to the surface roughness of the films.

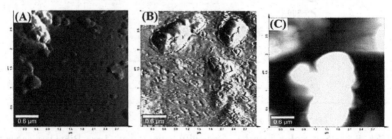

FIGURE 8.3   Two-dimensional AFM images of (1−x) $BiFeO_3$–x$NaNbO_3$–PMMA composite film samples for (a) x=0, (b) x=0.1, (c) x=0.5.

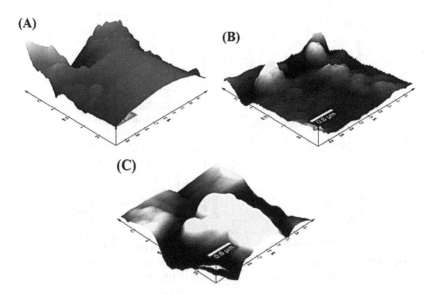

FIGURE 8.4 Three dimensional AFM images of $(1-x)$ $BiFeO_3$–$xNaNbO_3$–PMMA composite film samples for (a) $x=0$, (b) $x=0.1$, (c) $x=0.5$.

8.5.5 CONFOCAL RAMAN SPECTROSCOPY

Figure 8.5 shows Raman spectra of the samples. The backbone consists of polyethylene chain with H_3C—O—$\overset{\overset{O}{\|}}{C}$—$O$— and H_3C— attached to every second carbon atom. In the Raman spectra of pure PMMA, the bands of PMMA are visible. When nanoparticles were incorporated, the intensity of the peaks decreases and peaks corresponding to the nanoparticles also appear. The asymmetric stretches and the symmetric stretches of the methyl (CH_3) group in one structural repeat unit have been noted. The Raman frequency at 2950/cm is assigned to CH_3 symmetric stretching, 1729/cm is assigned to $C=O$ stretching, 1452/cm is assigned to CH_2 deformation, and 1242/cm is assigned to C–C degenerate stretching of CCl_4. The wagging mode of methylene group is observed at 1065/cm. The band at 814/cm is due to symmetric CCl_4 stretching and the band at 601/cm represents O–C=O deformation. The bands at 484/cm and 365/cm were assigned to C–C skeletal deformation of CCl_4.[35]

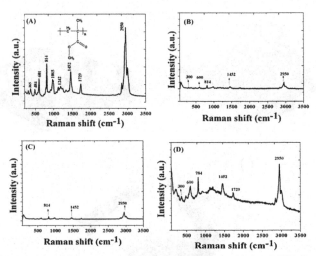

FIGURE 8.5 Confocal Raman spectra of (a) PMMA alone, (b) $(1-x)$ $BiFeO_3$–$xNaNbO_3$–PMMA composite film samples for (a) $x=0$, (b) $x=0.05$, (c) $x=0.1$, (d) $x=0.5$.

The small intensity bands present in Figure 8.5(b–d) are that of the ceramic particles. The modes which lie between 300 and 600/cm are characteristic modes of $BiFeO_3$.[36]

The Raman band in the range of 500–700/cm (612/cm) and 150–300/cm (251/cm) are assigned to the Nb–O–Nb vibrations of the NbO_6 octahedrons present in the crystalline structure of $NaNbO_3$.[37–40]

Raman images corresponding to the Raman spectra of $x=0, 0.1, 0.5$ compositions are shown in Figure 8.6. The dark blue region represents the nanoparticles and the light blue region is the polymer matrix.

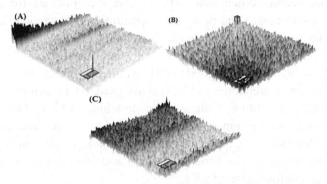

FIGURE 8.6 Raman images of $(1-x)$ $BiFeO_3$–$xNaNbO3$–PMMA composite film samples for (a) $x=0$, (b) $x=0.05$, (c) $x=0.1$, (d) $x=0.5$.

8.5.6 ELECTRICAL AND MAGNETIC STUDY OF THE FILM SAMPLES

Figure 8.7 shows dielectric constant as a function of frequency of $BiFeO_3$–$NaNbO_3$ ceramics at room temperature. The loss tangent as a function of frequency is plotted in Figure 8.8. Figures 8.7 and 8.8 show that dielectric constant and loss of these samples displays a slightly descending trend with increasing frequency from 10 to 2 MHz. The low loss tangent may be attributed to the low leakage current which may be due to the polymer PMMA which is accepted as a polymer with the low dielectric loss. The nanosize grains could also be another reason for the low loss. When the size of the grains is small, there will be large insulating boundaries between the grains, which act as barriers for current conduction, leading to low leakage current. The maximum value of the dielectric constant observed is 87 while the loss tangent is very low.[39,41]

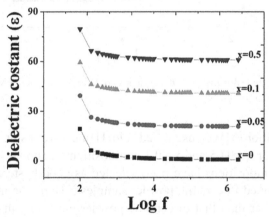

FIGURE 8.7 (See color insert.) Dielectric constant of $BiFeO_3$–$NaNbO_3$–PMMA composite film as a function of frequency.

Compared to the previous reports of BFO, no enhancement in the dielectric constant is observed due to the addition of PMMA.[42] Compared to the study on $BiFeO_3$–$NaNbO_3$ ceramic powder[36] and P (VDF-trifluoro ethylene [TrFE]) ceramic film, the current material possess low dielectric property. In some early studies of PMMA, the dielectric constant is reported to be less than four.[43] Here, the increased value of dielectric constant compared to other PMMA composite film could be due to the contribution of the ceramic particle, mainly the $NaNbO_3$ phase. The observed value of dielectric constant of the polymer–ceramic composite may be due to the interfacial polarization among ceramic–polymer interfaces. The dielectric constant (ε) and loss tangent (tan δ) were found to increase with increase in the amount of $NaNbO_3$.

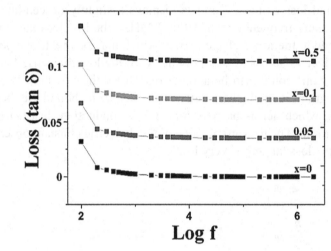

FIGURE 8.8   (See color insert.) Dielectric loss of $BiFeO_3$–$NaNbO_3$–PMMA composite film as a function of frequency.

Magnetization (M) versus applied field (H) measured using VSM at room temperature for $x = 0.05$ and $x = 0.1$ compositions is shown in Figure 8.9. These two compositions have been selected as good hysteresis behavior could be obtained in ceramic powder samples.[20] Here, the magnetization values are lower than that of ceramic powder indicating that PMMA do not alter the magnetic property for the current samples. The values are found to be much higher than that of the bulk BFO and BFO composites prepared by other techniques. Here, the remanent magnetization and coercivity (Hc) of both samples are same and the maximum magnetization value attained is 0.043 emu/g. The magnetization curves are not saturated within the field of 30 k·Oe and one could observe ferromagnetic behavior for both samples at room temperature. Recent researches on nanoparticles of BFO shows ferromagnetism at room temperature.[44]

Moreover, numerous antiferromagnetic materials in bulk form turns to ferromagnetic when the size is reduced to the nanometer scale.

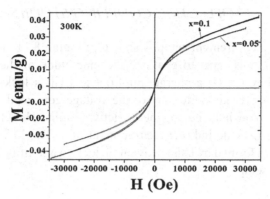

FIGURE 8.9 Magnetization (M) versus magnetic field (H) plot of x = 0.05 and x = 0.1 compositions at 300 K.

Temperature-dependent magnetizations for the (1−x) BFO–xNaNbO₃– PMMA films were measured during zero field cooled (ZFC) and field cooled (FC) processes with H = 200 Oe, as shown in Figure 8.10. The curves show a small cusp at 50 K which is common in BFO. For bulk BFO single crystals, this transition has been reported at 50 K as a very sharp Cusp.[45] But for nanoparticles the peak becomes broad. Moreover, the ZFC and FC curves diverge at low temperature. This divergence between ZFC and FC magnetization curves is similar to that found for other ferro and ferrimagnetic materials. The unsaturated hysteresis loop and low-temperature splitting of FC–ZFC curve is the characteristics observed in spin glasses.[46]

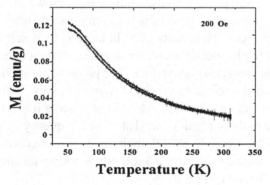

FIGURE 8.10 Temperature-dependent magnetizations measurement at 200 Oe for x = 0.1 composition.

8.5.7 MAGNETOELECTRIC COUPLING STUDIES

ME coupling measurement setup is shown in Figure 8.11. There are two methods commonly used to evaluate the magnitude of the (ME) coefficient (α). They are (i) a charge amplifier and (ii) a lock-in amplifier. These methods provide a measure of the voltage generated as a result of the magnetic field-induced magnetostrictive strain. Therefore, both of these methods provide indirect measurement of α. The ME effect of our samples was performed by using a dynamic lock-in amplifier ME coupling measurement setup.

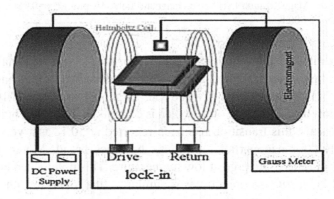

FIGURE 8.11 Schematic view of a magnetoelectric coupling measurement setup.

The induced polarization under an applied magnetic field was monitored by using a hall probe with the configuration shown in the figure. The DC bias is generated using a set of permanent magnets, while the AC field is produced by a set of Helmholtz coils. Induced voltage across the sample was measured by the lock-in amplifier in differential mode to subtract the common-mode induction contribution. The phase of the applied frequency was locked to pick pure-induced signal from the sample. The experiment involves the measurement of the induced voltage as a function of the variable amplitude of the applied AC field at a fixed frequency of 850 Hz which was produced by a couple of Helmholtz coils.[1,47–53] According to theory, the induced voltage should have a linear relationship with amplitude of the AC applied field and α is given by

$$\alpha = 1/t \ (dV/dH_{ac}) \tag{8.1}$$

where "V" is the induced voltage, "H_{ac}" is the amplitude of the applied AC magnetic field, "α" is the ME coupling coefficient which corresponds to induction of polarization by a magnetic field or of magnetization by an electric field and "t" is the thickness of the sample.

The ME voltage measured as a function of AC and DC magnetic fields are plotted in Figure 8.12. Here, the DC field has been fixed as 1000 Oe and the measurement was done at room temperature. A linear dependence of ME voltage to the applied magnetic field is observed. The same behavior is observed for BFO–NaNbO$_3$ ceramics[20] and BFO–NaNbO$_3$-P (VDF-TrFE) composite films. The ME coupling coefficient (α) is determined from the slope of the ME curve. A maximum value of coupling coefficient is observed for x = 0.1 composition which is 1.34/Vcm/Oe. By increasing x value from x = 0 to x = 0.1, the ME voltage increases. The multiferroic nature of x = 0.1 is higher compared to other compositions. This is mainly because of the ferroelectric polymer and NaNbO$_3$. However, further increase of x (x > 0.1) reduces the magnetic phase (i.e., BiFeO$_3$ content) and hence the multiferroic nature reduces.

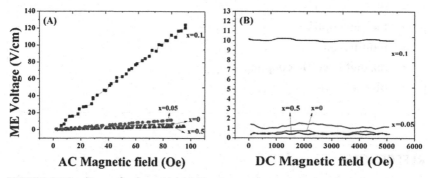

FIGURE 8.12 (See color insert.) (a) ME voltage measured as a function of AC magnetic field, (b) ME voltage measured as a function of DC magnetic field.

8.4 CONCLUSION

BiFeO$_3$–NaNbO$_3$–PMMA composites film samples were fabricated by solvent casting and fibers were fabricated by electrospinning. The multiferroic nature of the ceramic–polymer film composites were revealed by the ME coupling studies conducted at room temperature. Room temperature magnetization studies show hysteresis curves which represent weak ferromagnetic behavior. Temperature-dependent magnetization measurement reveals the spin glass behavior of the composites. Good electrical properties were observed by dielectric studies. The low loss tangent of the material is an interesting property. The ME coupling determined by a lock-in amplifier technique shows good coupling between the electric and magnetic phase for the composites. It is observed that addition of conducting polymer PMMA enhances the electric property. As opposed to bulk $BiFeO_3$, which demonstrates weak magnetization, the nanostructured $BiFeO_3$ in fiber form had shown room temperature ferromagnetism. Compared to film samples, fiber composites have shown reduced electric property.

KEYWORDS

- composite films
- composite nanoparticles
- dielectric study
- lock in amplifier
- multiferroics
- magnetoelectric coupling
- piezoelectric

REFERENCES

1. Arevalo Lopez, M.; Angel, J. Paul Attfield. *Phys. Rev. B* **2013**, *87*, 104416.
2. Eliseev, E. A.; Glinchuk, M. D.; Khist, V. V.; Lee, C. W.; Deo, C. S.; Behera, R. K.; Morozovska, A. N. New Multiferroics Based on EuxSr1−xTiO3 Nanotubes and Nanowires. *J. Appl. Phys.* **2013**, *113*, 024107.
3. Yoon, J. R.; Han, J. W.; Lee, K. M.; Lee, H. Y. *Trans. Electr. Electron. Mater.* **2009**, *10*, 1229–7607.
4. Ahlawat, A.; Satapathy, S.; Bhartiya, S.; Singh, M. K.; Choudhary, R. J. BiFeO3/poly (Methyl Methacrylate) Nanocomposite Films: A Study on Magnetic and Dielectric Properties. *Appl. Phys. Lett.* **2014**, *104*, 042902.
5. Chappert, C.; Fert, A.; Van Dau, F. N. The Emergence of Spin Electronics in Data Storage. *Nat. Mater.* **2007**, *6*, 11, 813–823.

6. Heping Zhou. Multiferroic Ferrite/Perovskite Oxide Core/Shell Nanostructures. *J. Mater. Chem.* **2010**, *20*, 10665.

7. Wang, S. X.; Taratorin, A. M. *Magnetic Information Storage Technology*; Apple Academic Press: USA, 1999; Vol. 1, Ch3 and 6.

8. Scott, J. F. Data Storage. Multiferroic Memories. *Nat. Mater.* **2007**, *6*, 256–257.

9. Kimura, T.; Goto, T.; Shintani, H.; Ishizaka, K.; Arima, T.; Tokura, Y. Magnetic Control of Ferroelectric Polarization. *Nature* **2003**, *426*(6962), 55–58.

10. Bibes, M.; Barthelemy, A. Multiferroics: Towards a Magnetoelectric Memory. *Nat. Mater.* **2008**, *7*, 425–426.

11. Chen, X.; Hochstrat, A.; Borisov, P.; Kleemann, W. Magnetoelectric Exchange Bias Systems in Spintronics. *Appl. Phys. Lett.* **2006**, *89*(20), 202–508.

12. Scott, J. F. Folding Catastrophes Due to Viscosity in Multiferroic Domains: Implications for Room Temperature Multiferroic Switching. *J. Phys. Condens. Matter.* **2015**, *27*, 492001.

13. Zhuang, J.; Su, L. W.; Wu, H.; Bokov, A. A.; Liu, M.; Ren, W.; Ye, Z. G. Coexisting Ferroelectric and Magnetic Morphotropic Phase Boundaries in Dy-Modified BiFeO3-PbTiO3 Multiferroics. *Appl. Phys. Lett.* **2015**, *107*, 182906.

14. Gajek, M.; Bibes, M.; Fusil, S. et al. Tunnel Junctions with Multiferroic Barriers. *Nat. Mater.* **2007**, *6*, 296–302.

15. Wang, J.; Neaton, J. B.; Zheng, H. et al. Epitaxial BiFeO3 Multiferroic thin Film Heterostructures. *Science* **2003**, *299*(5613), 1719–1722.

16. Zhao, T.; Scholl, A.; Zavaliche, F. et al. Electrical Control of Antiferromagnetic Domains in Multiferroic BiFeO3 Films at Room Temperature. *Nat. Mater.* **2006**, *5*(10), 823–829.

17. Jang, H. W.; Baek, S. H.; Ortiz, D. et al. Epitaxial (001) BiFeO3 Membranes with Substantially Reduced Fatigue and Leakage. *Appl. Phys. Lett.* **2008**, *92*(6), 062910.

18. Lou, X. J.; Yang, C. X.; Tang, T. A.; Lin, Y. Y.; Zhang, M.; Scott, J. F. Formation of Magnetite in Bismuth Ferrite Under Voltage Stressing. *Appl. Phys. Lett.* **2007**, *90*(26), 262908.

19. Takahashi, K.; Kida, N.; Tonouchi, M. Terahertz Radiation by an Ultrafast Spontaneous Polarization Modulation of Multiferroic BiFeO$_3$ Thin Films. *Phys. Rev. Lett.* **2006**, *96*(11), 1–4.

20. Ryzhii, V. Heterostructure Terahertz Devices. *J. Phys; Condens. Matter* **2008**, *20*(38), 380301.

21. Catalan, G.; Scott, J. F. Physics and Applications of Bismuth Ferrite. *Adv. Mater.* **2009**, *21*, 2463–2485.

22. Yi, Y.; Zhang, X.; Zhao, Y. G.; Jiang, N.; Yu, R.; Wang, J. W.; Fan, C.; Sun, X. F.; Zhu, J. Atomic-Scale Study of Topological Vortex-Like Domain Pattern in Multiferroic Hexagonal Manganites. *Appl. Phys. Lett.* **2013**, *3*, 32901.

23. Nan, C. W.; Bichurin, M. I.; Dong, S. X.; Viehland, D.; Srinivasan, G. Mulriferroic Magnetoelectric Composites: Historical Perspective, Status and Future Directions. *J. Appl. Phys.* **2008**, *103*, 31101.

24. Wang, B. Y.; Wang, H. T.; Shashi, B.; Singh, Y. C.; Shao, Y. F.; Wang, C. H.; Chuang, P. H.; Yeh, J. W.; Chiou, C. W.; Pao, H. M.; Tsai, H. J.; Lin, J. F.; Lee, C. Y.; Tsai, W. F.; Hsieh, M. H.; Tsaif, Pong, W. F Effect of geometry on the magnetic properties of CoFe2O 4-PbTiO3 multiferroic composites. *RSC Adv.* **2013**, *3*, 7884.

25. Raneesh, B.; Saha, A.; Das, D.; Sreekanth, P.; Philip, R. Nandakumar Kalarikkal. *RSC Adv.* **2015**, *5*, 12480.
26. Ummer, R. P.; Raneesh, B.; Thevenot, C.; Rouxel, D.; Thomas, S.; Kalarikkal, N. Electric, Magnetic, Piezoelectric and Magnetoelectric Studies of Phase Pure (BiFeO3–NaNbO3)–(P(VDF-TrFE)) Nanocomposite Films Prepared by Spin Coating. *RSC Adv.* **2016**, *6*, 28069.
27. Jayakumar, O. D.; Mandal, B. P.; Majeed, J.; Lawes, G.; Naik, R.; Tyagi, A. K. Inorganic–Organic Multiferroic Hybrid Films of Fe3O4 and PVDF with Significant Magneto-Dielectric Coupling. *J. Mater. Chem. C* **2013**, *1*, 3710.
28. Stingaciu, M.; Reuvekamp, P. G.; Tai, C. W.; Kremer, R. K.; Johnson, M. The magnetodielectric effect in BaTiO3–SrFe12O19 nanocomposites. *J. Mater. Chem. C* **2014**, *2*, 325–330.
29. Binek, C. Controlling Magnetism with a Flip of a Switch. *Physics* **2013**, *6*, 13.
30. Lawes, G.; Srinivasan, G. Introduction to magnetoelectric coupling and multiferroic films. *J. Phys. D Appl. Phys.* **2011**, *44*, 243001.
31. Martin, L. W.; Chu, Y. H.; Ramesh, R. Advances in the Growth and Characterization of Magnetic, Ferroelectric, and Multiferroic Oxide Thin Films. *Mater. Sci. Eng. R* **2010**, *68*, 89–133.
32. Ma, J.; Hu, J.; Li, Z.; Nan, C. W. Recent Progress in Multiferroic Magnetoelectric Composites: from Bulk to Thin Films. *Adv. Mater.* **2011**, *23*, 1062–1087.
33. Martin, L. W.; Crane, S. P.; Chu, Y. H.; Holcomb, M. B.; Gajek, M.; Huijben, M.; Yang, C. H.; Balke, N.; Ramesh, R. Multiferroics and Magnetoelectrics: Thin Films and Nanostructures *J. Phys. Condens. Matter* **2008**, *20*, 434220.
34. Yoon, J. R.; Han, J. W.; Lee, K. M.; Lee, H. Y. *Trans. Electr. Electron. Mater.* **2009**, *10*, 1229–7607.
35. Sun, W. H.; Hung, D. S.; Song, T. T.; Fu, Y. P.; Lee, S. F. Microwave Shielding Characteristics of PMMA/BiFeO3 Composites. *IEEE Trans. Magn.* **2011**, *47*, 4306–4309.
36. Ummer, R. P.; Sreekanth, P.; Raneesh, B.; Philip, R.; Rouxel, D.; Thomas, S.; Kalarikkal, N. Electric, Magnetic and Optical Limiting (short pulse and ultrafast) Studies in Phase Pure (1 − x)BiFeO3–xNaNbO3 Multiferroic Nanocomposite Synthesized by the Pechini Method. *RSC Adv.* **2015**, *5*, 67157.
37. Matsushita, A.; Ren, Y.; Matsukawa, K.; Inoue, H.; Minami, Y.; Noda, I.; Ozaki, Y. Two-Dimensional Fourier-Transform Raman and Near-Infrared Correlation Spectroscopy Studies of Poly(methyl methacrylate) Blends: 1. Immiscible Blends of Poly(methyl methacrylate) and Atactic Polystyrene. *Vib. Spectrosc.* **2000**, *24*, 171–180.
38. Pradhan, S. K. Raman and Electrical Studies of Multiferroic BiFeO. *J. Mater Sci. Mater Electron.* **2013**, *24*, 3581–3586.
39. Heracleous, E.; Lemonidou, A. A. Ni–Nb–O Mixed Oxides as Highly Active and Selective Catalysts for Ethene Production via Ethane Oxidative Dehydrogenation. Part I: Characterization and Catalytic Perf. *J. Catal.* **2006**, *237*, 162–174.
40. Wang, X. B.; Shen, Z. X.; Hu, Z. P.; Qin, L.; Tang, S. H.; Kuok, M. H. High Temperature Raman Study of Phase Transitions in Antiferroelectric NaNbO3. *J. Mol. Struct.* **1996**, *385*, 1–6.

41. Kumar, M. M.; Palkar, V. R.; Srinivas, K.; Suryanarayana, S. V. Ferroelectricity in a Pure BiFeO3 Ceramic. *Appl. Phys. Lett.* **2000**, *76*, 2764.
42. Huang, B. X.; Wang, K.; Church, J. S.; Li, Y. S. Characterization of Oxides on Niobium by Raman and Infrared Spectroscopy. *Electrochim. Acta* **1999**, *44*, 2571–2577.
43. Pradhan, A. K.; Zhang, K.; Hunter, D.; Dadson, J. B.; Loutts, G. B.; Bhattacharya, P.; Katiyar, R.; Zhang, J.; Sellmyer, D. J.; Roy, U. N.; Cui, Y.; Burger, A. Magnetic and Electrical Properties of Single-Phase Multiferroic. BiFeO3. *J. Appl. Phys.* **2005**, *97*, 90–93.
44. Mahesh Kumar, M.; Palkar, V. R. Ferroelectricity in a pure BiFeO3 ceramic. *Appl. Phys. Lett.* **2000**, *76*, 2764.
45. Moonkyong, N.; Shi-Woo, R. Electronic characterization of Al/PMMA [poly (methyl methacrylate)] p-Si and Al/CEP (cyanoethyl pullulan)/p-Si structures. *Org. Electron.* **2006**, *7*, 205–212.
46. Lan T., Kaviratna P. D, Pinnavaia T., J. On the nature of polyimide-clay hybrid composites. *Chem. Mater.* **1994**, *6*, 573.
47. Eliseev, E. A.; Glinchuk, M. D.; Khist, V. V.; Lee, C. W.; Deo, C. S.; Behera, R. K.; Morozovska, A. N. New Multiferroics Based on Eu x Sr 1−x TiO 3 Nanotubes and Nanowires. *J. Appl. Phys.* **2013**, *113*, 24107.
48. Suastiyanti, D.; Soegijono, B.; Hikam, M. Magnetic Behaviors of BaTiO3-BaFe12O19 Nanocomposite Prepared by Sol-Gel Process Based on Differences in Volume Fraction. *Adv. Mater. Res.* **2013**, *789*, 118–123.
49. Giordano, S.; Dusch, Y.; Tiercelin, N.; Pernod, P.; Preobrazhensky, V. Thermal Effects in Magnetoelectric Memories with Stress-Mediated Switching *J. Phys. D: Appl. Phys.* **2013**, *46*, 325002.
50. Gerken, M. *AIP Adv.* **2013**, 062115.
51. Spaldin, N. A.; Fechner, M.; Bousquet, E.; Balatsky, A.; Nordstrom, L. Monopole-Based Formalism for the Diagonal Magnetoelectric Response. *Phys. Rev. B* **2013**, *88*, 094429.
52. Rocquefelte, X.; Schwarz, K.; Blaha, P.; Kumar, S.; van den Brink, J. High-Pressure Cupric Oxide: a Room-Temperature Spin-Spiral Multiferroic. *Nat. Comm.* **2013**, *4*, 2511.
53. Duraffourg, L.; Arcamone, J. *Nsanoelectromechanical Systems*; Wiley publishers: London, 2015, Vol. 1, Ch 2 and 3.

CHAPTER 9

CHLOROFLUOROCARBONS, OZONE LEVELS EVALUATION, PEOPLE ACTIONS, TROPOSPHERIC O$_3$, AND AIR QUALITY

FRANCISCO TORRENS[1,*] and GLORIA CASTELLANO[2]

[1]*Institut Universitari de Ciència Molecular, Universitat de València, Edifici d'Instituts de Paterna, 22085, E-46071 València, Spain*

[2]*Departamento de Ciencias Experimentales y Matemáticas, Facultad de Veterinaria y Ciencias Experimentales, Universidad Católica de Valencia San Vicente Mártir, Guillem de Castro, 94, E-46001 València, Spain*

Corresponding author. E-mail: torrens@uv.es

ABSTRACT

Ozone is a gas with a high oxidant power, which when present in high concentrations presents adverse effects on human health, vegetation, and materials. Ozone concentrates in two atmospheric strata: in the troposphere, it acts as a pollutant, while in the stratosphere, it is beneficial. In persons exposed to high ozone concentrations, it can produce damages mainly centered on the respiratory tract. Equipment that uses a normalized method are utilized to measure ozone environmental concentration, and allows performing continuous and automatic measurements of air ozone concentration. Ozone can reach high concentrations, mainly in rural sites located in the inland.

9.1 INTRODUCTION

In the laboratory, Figueruelo reviewed physical chemistry of the environment and environmental processes.[1–3] Rivera, Sapiña, Picó, and Escrivà through their research work made people aware about the keys to understand and stop Earth heating and climate change.[4–8]

In an earlier publication, converting cellulose, sucrose, and fructose into biofuel and new biofuels was discussed.[9–11] Seagrass (*Posidonia oceanica*) was argued as an indicator of seawater quality.[12] In this chapter, refrigerant chlorofluorocarbons (CFCs) are analyzed as causative agents of the *ozone (O_3) layer hole* and *greenhouse effect*. Ozone level evaluation allowed asking why people should act. Tropospheric O_3 permitted proposing questions (Qs) and answers (As) on air quality.

9.2 CHLOROFLUOROCARBONS

The refrigerator is not more than a century in people's houses, but it forms part of their daily life to such a point that not to have it seems inconceivable.[13] One can drink a glass of cold milk whenever he wants, and this soft murmur box in the kitchen corner inspired culinary masterworks, for example, refrigerator chocolate cake. In 2012, Royal Society proclaimed that the refrigerator was the most important invention in food history. Although, without doubt, it is a relief not to have to replenish one's pantry every 2 days, there always exists the possibility of discovering something disgusting lying in wait in the refrigerator background. Was it the case of a continent-sized *hole* in O_3 *layer* instead of some lettuce leaves? Nowadays, people know that the gases responsible for O_3 *layer hole* are CFCs, which refrigerants developed so as to replace the toxic gases that were used in refrigerators at the beginning of 20th century. Chlorinated CFCs demean with sunlight and release in the atmosphere dangerous radicals without Cl. Before CFCs, refrigerator manufacturers used chloromethane (CH_3Cl), ammonia and sulfur dioxide, which are all dangerous if they are inhaled in a closed environment. A refrigerant leak could be fatal.

9.2.1 A GOOD SOLUTION

Many historians cite a fatal explosion in which CH_3Cl was involved in a Cleveland (Ohio) hospital in 1929, as motivation for the development of nontoxic refrigerant gases. About 120 persons died in the disaster due to inhaling carbon monoxide (CO) and nitrogen oxides (NO_x), which were generated when burning down some X-ray films, and not by CH_3Cl. Anyway, chemical industry was already fully conscious of the dangers of using poisonous gases as refrigerants and searched for a solution. The year before Cleveland accident, Thomas Midgley, Jr., a General Motors researcher, synthesized a nontoxic halogens compound called dichlorodifluoromethane (CCl_2F_2, freon), which was the first CFC, although it was not utilized since the 1930s. Midgley's boss, Charles Kettering, was searching for the refrigerants that *were not inflammable and had no pernicious effects for people.* Today, it seems a bad omen that it was Midgley, who had just discovered tetraethyl lead (pollutant of petrol, used as antiknock agent), who received the job. In 1947, 3 years after Midgley's death, probably by suicide, Kettering wrote that freon just met the necessary criteria. It was not highly inflammable and *lacked pernicious effects for man and animals,* which was certain in a meaning: it caused no direct damage to persons and animals. Kettering noticed that no laboratory animal used in tests showed the least upset indication after inhaling freon. Midgley even demonstrated that CFCs were safe by inhaling them himself in a presentation. Therefore, CFCs were accepted as refrigerants. Midgley, who died early, could never understand the impact of his researches.

9.2.2 COVERING THE HOLE

In 1974, at the time as refrigerators and freezers were filled with Black Forest pies and Arctic rolls, the first CFC effect indication appeared in an article published by Sherry Rowland and Mario Molina, two chemists at the University of California. The report stated that O_3 *layer,* which filters the most harmful ultraviolet (UV) radiation that comes from the Sun, could be halved before mid-21st century if CFCs were not banned. It is not strange that the chemical companies that made money with refrigerants received the statements with concern. At that time, it was not demonstrated that CFCs caused harm to O_3 layer. Rowland and Molina described only a mechanism. Many people were already skeptical before the idea and warned that the economic consequences of banning CFCs would be grave. Another decade passed before irrefutable proofs of O_3 *layer hole* were available. British Antarctic Survey was monitoring O_3 in the atmosphere over Antarctic since toward the end of the 1950s and, in 1985, scientists had enough data to know that levels were decaying. Satellite data showed that the *hole* extended throughout Antarctic continent. Within 2 years, countries all over the world ratified Montreal Protocol relative to the substances that utilize O_3 *layer*, which established a calendar for CFC removal. In addition, nowadays, what lies in wait in the refrigerator background? Some manufacturers replaced CFCs with hydrofluorocarbons (HFCs). As Cl is considered to be harmful, HFCs are a common substitute. However, in 2012, Mario Molina was one of the authors of an article that manifested another problem: perhaps HFCs do not damage O_3 *layer*, but some are *greenhouse effect* gases, 1000 times more powerful than carbon dioxide (CO_2). In July 2014, in fifth consecutive year, Montreal Protocol signers discussed its extension to HFCs.

9.3　EVALUATION OF THE LEVELS OF OZONE

Ozone is a reactive molecule that continues reacting with other pollutants present in the air, forming a set of several tens of different substances, for example, peroxyacyl nitrates (PANs), hydrogen peroxide (H_2O_2), hydroxyl radical ($\cdot OH$), and formaldehyde, which as a whole can significantly damage plants, and can cause eye irritation and respiratory problems.[14] The type of pollution worsens in hot zones and with a high rate of sun radiation. Spring-summer is the most favorable period for the formation of the type of pollution although, in winter, thermal-inversions development and lower mixing-layer development favor gaseous and particulate pollutants concentration in lower atmospheric layers, being able to aggravate the problem in certain scenes as they hinder air renewal and pollutants removal. Although O_3 precursors are emitted in urban and industrial zones, O_3 impact is registered in rural zones and of regional background surrounding emitting focuses, while in urban environments, O_3 is consumed to oxidize primary NO, resulting in low O_3 levels.

9.4　WHY SHOULD PEOPLE ACT?

Epidemiology played a crucial role in identifying the threat that atmospheric pollution represents for health. Epidemiological studies provide key information to establish air quality standards that should be achieved to protect population health. In Europe, World Health Organization estimates that the high concentrations of particles in suspension are associated with 300,000 annual premature deaths, which means, on average, the life expectancy of Europeans decay by at least 1 year. Scientific evidence allows concluding that a rise in particulate matter levels leads to excesses of mortality risk, especially by cardiorespiratory cause, respiratory and cardiac disease hospitalizations, asthma decline, persistent respiratory symptoms, lung function alteration, and disability. Even if all the biomechanisms of the particles are not known, epidemiological studies results point to them as an important health risk factor for the short and long terms. Children, old persons, and people that suffer chronic respiratory or cardiovascular diseases are most sensitive to atmospheric pollution effects. The health impact and high economic cost due to air pollutants present in the environment justify that solid measures should be taken for atmospheric pollution reduction. Intervention studies are useful to quantify health benefits that air quality regulations and emission control measures cause. In the short term, the decay of air pollutant levels supposes improvements in asthmatic population health. In the long term, positive health impact is even greater as the trend of the annual average of the deaths by all and specific causes decreases and increases life expectancy. The available information on the effects that atmospheric pollution causes to health is important enough to act and decrease pollution levels in cities. The interventions that involve an air quality improvement are accompanied by substantial and appreciable benefits in public health terms. It was proved that the adequate management of air quality improves population health as the reduction of pollutant concentration levels is associated with morbidity and mortality decays by all and, in particular, respiratory and cardiovascular causes. Atmospheric pollution is a serious motive of health worry, but real action possibilities exist if people rely on firm political will and civic conscience. Measures guided to obtain a cleaner and healthier environment are essential to have a really sustainable future.

9.5   TROPOSPHERIC OZONE AND AIR QUALITY

Mantilla et al. proposed some questions and answers on tropospheric O_3 and air quality.[15]

Q1. Is O_3 a dangerous pollutant?

A1. Ozone is a gas with a high oxidant power, which when present in high concentrations presents adverse effects on human health, vegetation, and materials. The incidence is more evident in short-time concentration acute episodes than in the case of expositions to moderate levels during long periods. It is necessary to prevent the exposition of children, old persons, and individuals with respiratory problems and, especially, in physical activities, which rise ventilation.

Q2. What is the origin of O_3 in the atmosphere?

A2. Ozone concentrates in two atmospheric strata: in the lowest layer (troposphere), it acts as a pollutant (*bad* O_3), which is formed due to reactions with the substances emitted by traffic, industry, and vegetation intervene. In the stratosphere (12–50 km), O_3 is found in higher concentrations and is originated from solar radiation; its effect is beneficial, being responsible for filtering UV-B radiation, which is harmful to living beings before it reaches the Earth surface (*good* O_3).

Q3. How does O_3 affect environment and health?

A3. In persons exposed to high concentrations, it can produce damages mainly centered on the respiratory tract. In plants, it can cause a reduction in growing measures of leaves, trunk, roots, and fruits. In materials, O_3 favors and powers corrosion processes, speeding up, in general, their degradation.

Q4. How is O_3 environmental concentration measured?

A4. Equipment that use a normalized method are utilized, which allow performing continuous and automatic measurements of air O_3 concentration. Generalitat Valenciana relies on a vigilance net distributed throughout Valencia Community territory, which offers information on O_3 and other pollutants concentrations, and weather variables, making possible a rapid and reliable diagnostic of air quality state.

Q5. How is the problem of the pollution by O_3 tackled?

A5. Different European Directives were enacted, transposed to the Spanish legislation, which establish the exigencies for the action by administration versus O_3 pollution. The rules establish target values of O_3 concentrations to protect human health and vegetation, and population information and warning requisites.

Q6. What is the present situation of Valencia Community with regards to the pollution by O_3?

A6. Ozone can reach high concentrations, mainly in rural sites located in the inland. In some stations, health protection threshold defined in Directive 2002/3/EC goes beyond during 20 days per month in summer, while Directive recommends that it must not go beyond more than 25 days per year. On the contrary, information threshold (defined at 180 $\mu g/m^3$ as hourly average) does not usually go beyond more than 6 days per year, and in concentrations not greater than 20 $\mu g/m^3$ over the threshold. The warning threshold was not reached in any measurement point in the Valencia Net of Atmospheric-Pollution Vigilance and Control.

Q7. What does the vigilance program of tropospheric O_3 (PREVIOZONO) developed by Valencia Community consists of?

A7. Generalitat Valenciana, developed since 1999, in collaboration with Centre of Environmental Studies of Mediterranean Foundation, a vigilance program of the concentration levels of tropospheric O_3, formed and carried out legal exigencies relative to population information and warning in situations of overcoming of thresholds established on this matter, at the same time, continued vigilance and information is kept independent of the occurrence or not of overcoming episodes. When a high probability of registering an information and/or warning threshold overcoming occurs or is foreseen, a specific plan is activated with the purpose of securing information to population, starting the distribution channel through Civil Defence protocols. A messaging service to mobiles is available with information about the occurrences of information and/or warning threshold overcoming (station, hour, concentration, duration).

Q8. How can the pollution by O_3 be fought?

A8. Ozone concentration reduction, as a secondary pollutant, should be necessarily performed by limitating the emission of its precursors, mainly NO_x and hydrocarbons.

ACKNOWLEDGMENT

The authors thank support from Generalitat Valenciana (Project No. PROMETEO/2016/094) and Universidad Católica de Valencia *San Vicente Mártir* (Project No. UCV.PRO.17-18.AIV.03).

KEYWORDS

- Earth heating
- climate change
- toxic refrigerant gas
- freon
- ozone layer hole
- greenhouse effect gas
- seasonal air pollutant
- particulate matter
- epidemiological study
- air pollution reduction

REFERENCES

1. Figueruelo, J. E.; Dávila, M. M. Química Física del Medio Ambiente; Reverté: México, 2001.
2. Figueruelo, J. E.; Dávila, M. M. Química Física del Ambiente y de los Procesos Medioambientales; Reverté: Barcelona, 2004.
3. Figueruelo, J. E.; León, L. M. Introducción a la Química-Física para las Ciencias Ambientales; Universidad del País Vasco: Bilbao, 2011.
4. Rivera, A. El Cambio Climático: El Calentamiento de la Tierra; Debate: Madrid, 2000.
5. Sapiña, F. Un Futur Sostenible? El Canvi Global Vist per un Químic Preocupat; Sense Fronteres No. 10, Universitat de València–Bromera: Alzira (València), 2001.
6. Sapiña, F. El Repte Energètic: Gestionant el Llegat de Prometeu; Sense Fronteres No. 20, Universitat de València–Bromera: Alzira, València, Spain, 2005.
7. Picó, M. J. El Canvi Climàtic a Casa Nostra; Actual No. 9, Bromera: Alzira, València, Spain, 2007.
8. Escrivà, A. Encara No És Tard: Claus per a Entendre i Aturar el Canvi Climàtic; Sense Fronteres No. 40, Universitat de València–Bromera: Alzira, València, Spain, 2017.
9. Torrens, F.; Castellano, G. Book of Abstracts, 6th International Symposium on Ozone Applications, Havana, Cuba, June 28–July 1, 2010; The Ozone Research Center of Cuba: Havana, Cuba, 2010, P-1.
10. Torrens, F.; Castellano, G. Book of Abstracts, V International Symposium on Environment, Havana, Cuba, June 28–July 1, 2010; The Ozone Research Center of Cuba: Havana, Cuba, 2010, P-1.

11. Torrens, F.; Castellano, G. *Book of Abstracts, II Ozone Therapy International Congress from FIOOT*, Havana, Cuba, June 28–July 1, 2010; The Ozone Research Center of Cuba: Havana, Cuba, 2010; P-1.
12. Castellano, G.; Tena, J.; Torrens, F. Classification of Polyphenolic Compounds by Chemical Structural Indicators and its Relation to Antioxidant Properties of *Posidonia oceanica* (L.) Delile. MATCH Commun. *Math. Comput. Chem.* **2012**, *67*, 231–250.
13. Birch, H. 50 Chemistry Ideas You Really Need to Know; Quercus: London, 2015.
14. Querol, X.; Viana, M.; Moreno, T.; Alastuey, A. eds. Bases Científico-técnicas para un Plan Nacional de Mejora de la Calidad del Aire; Informes CSIC No. 3, Consejo Superior de Investigaciones Científicas: Madrid, 2012.
15. Mantilla, E.; Castell, N.; Salvador, R.; Azorín, C.; Millán, M.; Miró, J. V.; Juan, L. Ozono Troposférico y Calidad del Aire; Generalitat Valenciana: València, 2017.

PART II
Chemical Informatics

LIGNIN UTILIZATION AS A WOOD SURFACE COATING AND ITS ROLE IN FEASIBILITY ASSESSMENT OF LIGNOCELLULOSE-BASED BIOETHANOL IN INDONESIA: A CHEMINFORMATICS APPROACH

TEUKU BEUNA BARDANT[1], HERU SUSANTO[1,2,*], and ARIEF AMIER RAHMAN SETIAWAN[1]

[1]Indonesian Institute of Science, Jakarta Timur, Indonesia

[2]Department of Information Management, College of Management, Tunghai University, Taichung, Taiwan

*Corresponding author. E-mail: heru.susanto@lipi.go.id

ABSTRACT

The study emphasized on palm plantation biomass waste because it is the most concern potency for bioethanol raw material in Indonesia. Here, Indonesia is the world leading producer of palm oil, as consequence number of palm oil plantation was developed since 1968. Only 79.209 Ha belongs to the Indonesian government and the other 40.451 Ha runs by private. Empty fruit bunch (EFB) produced from palm oil production since the EFB composition in fresh fruit bunch is similar to palm oil, 22–23%w. This huge potency was waiting to be explored for the sake of Indonesian prosperity. The most efficient way to detach palm fruit from it bunch is by steam cooking the fresh bunch in the autoclave. Palm plantation biomass waste is used as raw material for ethanol production. To build self-sufficient energy of integrated production plant which produces bioethanol, lignin,

lignin derivatives, and wood coating agent, cheminformatics approach was suitable for preliminary analysis in planning plant intensification or product diversification. Several analyses from the previous study in transforming old pulp mill into biorefinery plant with ethanol production can be applied in a model of bioethanol production plant from palm oil EFB that was planned to be built in Indonesia. Information and research results from the existing bioethanol pilot plant which gave significant support in developing cheminformatics preliminary analysis. Production of lignin derivatives as an integral part of bioethanol production plant increase the overall energy efficiency which leads to the increasing feasibility of integration process.

10.1 INTRODUCTION

Bioethanol production from cellulose-based material is one of the popular bioprocess engineering research fields in Indonesia in the last 5 years. Palm plantation biomass waste became the most concern potency for bioethanol raw material since Indonesia is the world leading producer of palm oil. Indonesia palm oil plantation was developed since 1968 from only 79.209 Ha which belongs to the Indonesian government and the other 40.451 Ha plantation was run by private companies. The smallholder industrialist started to contribute in producing palm oil in 1979, mainly as the impact of transmigration program created by the government which trigged agroindustries outside Java island. Along with the plantation expansion as shown in Figure 10.1, the production was kept increasing. In 2015, Indonesian total palm oil production was estimated to be 30,948,931 t of which only 2.49% came from government plantation.

The EFB (empty fruit bunch) potency can be extrapolated from palm oil production since the EFB composition in fresh fruit bunch is similar to palm oil, 22–23%w (Pahan, 2006). This huge potency was waiting to be explored for the sake of Indonesian prosperity. The most efficient way to detach palm fruit from it bunch is by the steam cooking fresh bunch in the autoclave. This process caused the EFB were available in wet conditions which became the main challenge for utilizing it as construction material. Recent utilization of EFB was as in situ composting materials which consume transportation cost from palm oil mill to the plantations. The utilization as compost was usually limited to the plantations having the same owner as that of the palm oil mill.

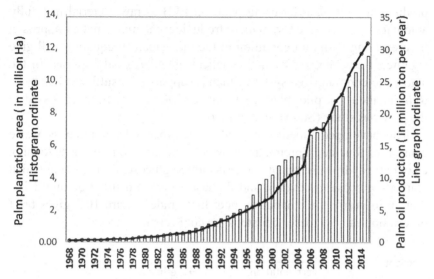

FIGURE 10.1 Indonesian palm oil plantation and palm oil production in the last five decades.

Bioethanol production from EFB next to palm oil mill gave a promising profitable solution. EFB as the wet raw material is not a problem since the pulping process is boiled in soda solution or other thermomechanical process. Energy produced from kernel and fiber combustion exceed the requirement of extraction plants which installing the latest technology. Energy assessment in seven different palm oil mills with capacity between 20 and 54 t/h had already conducted by using cheminformatics. It was concluded that total potential exported energy, in term of electricity, is ranging from 113 up to 902 MW (Nasrin et al., 2011). This excess energy can be used by bioethanol production in pulping and distillation unit. Recently, some mills use the excess energy for generating power plant and supply electricity to the adjacent dwellings. This utilization was limited by the wired infrastructure for distributions since the plant and planta-tion were usually placed in the relatively remote area. Storing the excess energy in form of bioethanol will give cheaper investment in distribution infrastructure by using trucking system.

Utilization of palm plantation biomass waste as raw material for ethanol production is still competitive and have been strongly supported by the Indonesian government. One of the significant supports is developing pilot plant in PP Kimia LIPI. Currently, Indonesia had established ethanol production pilot plant by using palm oil EFB as raw material in a fully automatic-computerized system. Introducing cheminformatics approach by presenting optimum condition in the mathematical equation will give significant advantage. The mathematical model is easily applied in the automatic-computerized system which is supported by sufficient database. Reports about the pilot plant performance had already been discussed in the previous study (Sudiyani et al., 2013).

The developed bioethanol production process from biomass waste consists of four steps, pretreatment which converts biomass waste to pulp, enzymatic hydrolysis which converts pulp to glucose, fermentation which converts glucose to ethanol, and the last steps are distillation and dehydration to purify the ethanol to meet fuel grade. Figure 10.2 gives brief description of how these four steps are connected.

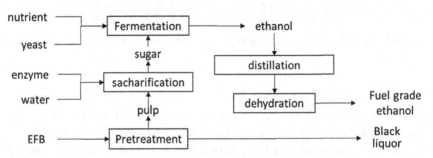

FIGURE 10.2 Block diagram of bioethanol production process from palm oil EFB.

Fermentation and distillation technology were also acknowledged since the history of humankind. Record and evidence of it can be found in every ancient civilization, that is, Egypt and China. This long-lived technology had constantly improved therefore nowadays, many industries are offering fermentation and distillation licenses. Thus, enzymatic hydrolysis is the unique and crucial steps in the whole process which determines the economic feasibility. The hydrolysis rate decreases during the process which leads to decreased yields and prolonged times. In the end, these will lead to higher production cost (Sadler, 1996; Bardant et al., 2010).

The closely relative benchmark for lingocellulose-based bioethanol production was starch-based bioethanol. In 1998, the portion of enzyme cost in ethanol production from starch in Brazil was up to 40% of the total cost, higher than raw material, utilities and other expenses which were 20, 28, and 12%, respectively. Through continuous research, the portion of enzyme cost became less than 8%. Although, there were contributions of increasing raw material prices and fuel which make their portion 32 and 35%, respectively. Research in developing high activity of amylase had reached the efficiency of 1 g of enzyme (protein equivalent) for 1 gal of ethanol. Kim (2007) observed that cellulase activity is 100 g for 1 gal of ethanol (Kim, 2007).

Novozyme became the leading industry in the race for providing reliable cellulase on the commercial scale. They claimed that their cellulase can be sold in US$2 per gallon which makes ethanol production from corncob feasible as it was produced from cassava. (novozymes press release Feb 15, 2010). Danisco. Inc as the competitor also introduced its latest cellulase in similar period. Reuters, Danisco Spokesman and Rene Tronborg claimed that by using their cellulase, bioethanol production from hays has feasibilities comparable with the production process that uses corn. This statement was announced in Renewable Fuels Association's 15th Annual National Ethanol Conference in Orlando, Florida. This event is a follow-up action of Barack Obama Statement to reduce dependency on fossil fuel at the early of 2010 (Kinney, 2012).

Other approach can be used in developing effective cellulose-to-ethanol process for liquid fuel application. To improve yields and reduce reduction time, and yet reducing the production cost, several technical approaches were studied. Conducting simultaneous hydrolysis and fermentation process for bioethanol production had already done by using the pulp of palm oil EFB (Triwahyuni et al., 2014; Triwahyuni et al., 2015). Kinetic study of enzymatic hydrolysis process had already conducted to synchronize the hydrolysis reaction rate with the fermentation rate (Barlianti et al., 2015). This study had intensively developed to a pilot-scale unit (Sudiyani et al., 2013). Conducting hydrolysis and fermentation simultaneously gave significant advantages compared to hydrolysis and fermentation in series (Dahnum et al., 2015).

There is huge opportunity to adapt existing pulping, fermentation and distillation technology, which will contribute to lower the investment cost and maintenance. Thus, easier to calculate depreciation which leads to lower the overall production cost. To adapt existing commercially reliable distillation unit for ethanol purification, hydrolysis and fermentation need to be designed for producing fermentation broth with the suitable specification. One of the well-known Indonesian ethanol producer, PT Madubaru Yogjakarta, produces ethanol from molasses and setting 8–9%v ethanol concentration in its feed specification for the distillation unit operating normally. The distillation unit itself was designed to purify fermentation broth which contains 5%v of ethanol at its lowest. The theoretical conversion of glucose to ethanol through fermentation using *Saccharomyces cerevisiae* is 51%. Thus, by also including cellulose content in pulp into account, the pulp loading in hydrolysis process need to be over 20%w of the reaction system. Surfactant addition proved to be a way to tackle the challenge of dealing with this high loading substrate. Several published report had explained the commercial surfactant addition in enzymatic hydrolysis of palm oil plantation waste in Indonesia (Bardant et al., 2012). Surfactant addition was able to increase cellulase performance not only for hydrolysis of EFB pulp but also other palm plantation waste such as pulp from palm oil trunk or even in water–hyacinth mechanical pulp (Bardant, 2013; Winarni, 2016; Bardant et al., 2017).

All aforementioned technological development in bioethanol production process from lignocellulose is still not enough to give competitive bioethanol price compared to gasoline. Best case operation of bioethanol pilot plant in Indonesia was in 2014 and gave bioethanol production cost of Rp. 10,500 (US $0.790). Indonesian nonsubsidized gasoline, with commercial brand PERTAMAX, had priced at Rp.11,200 (US $0.823) at that time. However, due to declining of crude oil price since 2014 to this recent time, it was lowering gasoline price in Indonesia to Rp. 6500 (US $0.489). Thus, significant effort needs to be conducted to increase the bioethanol feasibility. A potential approach was to utilize lignin side product as raw material of the high-added-value product. In harmony with surfactant addition approach, there was an effort to synthesize lignin derivative as amphipathic surfactant that is used for enhancing cellulase performance (Uraki, 2014). However, other lignin application was still required in larger amount.

Converting biomass waste into pulp is an ancient technology that defines civilization. Since Tsai Lun invented paper till now, the production technology had really evolved and established. This is a promising support for bioethanol production from biomass. A scope of improvement that had been explored mostly on increasing process efficiency by utilizing heat and chemicals as least as possible to meet next step requirement in bioethanol production process. In this chapter, lignin purification unit and lignin derivative synthesis were analyzed as part of the integrated system in producing bioethanol. But first in this following section, brief explanation of lignin purification process and lignin derivative synthesis were delivered. Moreover, performance comparison of resulted product to the commercial one in water protection properties were described to show the bright potential of lignin derivatives as wood surface coating agent.

10.2 LIGNIN DERIVATIVE APPLICATION AS WOOD SURFACE COATING AGENT

Purified lignin powder was prepared by using acid precipitation technology which is similar to the one that was assessed by Levasseur et al. (2011). The assessment of lignin purification unit will be discussed later in next section. The purification technology consists of three steps acidification, filtration, and washing until the washed water reaches to normal acidity. Lignin was then dried at 30°C in the oven. About 6.3 kg purified lignin powder with 6.33% moisture content which was obtained from 150 l black liquor as the side product of bioethanol pilot plant that used EFB as raw material. This lignin was the main ingredient for producing lignin derivatives.

The main idea in lignin derivative synthesis was to increase lignin solubility in organic solvent by reacting it with glycerol and ethylene glycol. The reaction process was conducted by using 1:1 reactant ratio at 150°C in basic condition. NaOH was used for creating basic condition and it was introduced by diluting it into glycerol or ethylene glycol. The concentrations as well as the reaction time were varied. The reaction was stopped by quenching it with cold 6 M H_2SO_4 solution and then extracted by using hexane. Unreacted lignin is precipitated at the bottom of aqueous phase. The resulted lignin derivatives lied in hexane phase is thus separated and dried. Lignin derivatives dry weight was measured and used for yield calculation.

From the production point of view, the experimental results that were shown in Figure 10.3 is very promising. Reacting lignin with glycerol showed nearly linear correlation with the concentration of NaOH that created the basic condition. Utilization of 2 M NaOH in glycerol could give 89%w yield of what further in this article will be called lignin–glycerol derivative. Different trend was found in reacting lignin with ethylene glycol. The resulted lignin–EG derivative was 75%w of reacted lignin with less reaction time. However, further analysis of the application of lignin–EG derivatives as wood surface coating agent in this study used lignin–EG that was obtained from 60 min reaction time due to its availability.

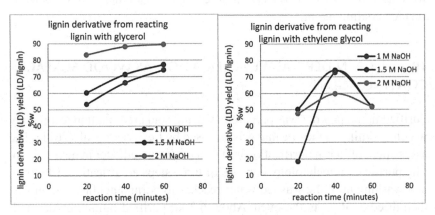

FIGURE 10.3   (See color insert.) Relation between lignin derivative yield to the reaction time with glycerol and ethylene glycol.

Commercial wood surface coating agent usually consists of two types of polymers that are applied as double layer coating. First, coating agent was applied for obtaining the desired wood color and then it becomes the bottom layer. The second coating agent covered the bottom layer to protect it from outer mechanical disturbance such as scratch or indentation. The second layer, which further in this chapter will be named as varnish layer, also enhances surface appearance by giving glossy effect. First, analysis of lignin derivatives application as wood coating agent was the full substitution of the bottom layer from wood commercial coating agent. Lignin derivatives were differentiated based on the reagent (lignin–glycerol and lignin–EG), and its basic condition where the lignin derivatives were synthesized.

Process for applications of lignin derivatives on the wood surface were conducted by diluting 2 g of each type of tested lignin derivatives into 10 ml commercial thinner. The resulted solution was spread onto prepared 4×4 cm plywood by using 1 cm brush. Brushing the plywood surface was repeated three to four times until all solution was used. All repetitions were conducted after the surface was completely dried by the previous brushing step. During the drying process, unattached lignin derivatives were swept away by surrounding air flow. After application of lignin derivatives, some plywood samples were then covered by varnish. At any point on the surface, it was covered by varnish solution only by one sweep of the brush. After the varnish layers were completely dried, the plywood pieces were weighted and the weight of the pieces was compared before coating process. The weight difference was then noted as mass deposition on the surface and presented in mg/cm^2. The results are shown in Figure 10.4.

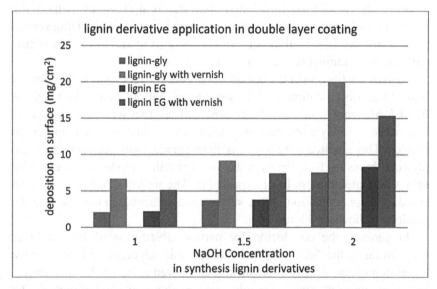

FIGURE 10.4 **(See color insert.)** The effect of NaOH concentration as catalyst in the synthesis to the amount of lignin derivative deposition on the wood surface when they were applied in double-layer coating.

The results showed that in any basic condition, the amount of lignin–EG derivatives that are attached to the wood surface are slightly higher than that of the lignin–glycerol and therefore seems to be insignificant. However, the concentration of NaOH for creating basic condition in synthesis reaction gave significant effect to the amount of lignin derivatives deposited on the wood surface. More lignin derivatives were attached to the wood surface if the synthesis reaction was carried in more basic condition.

Application of varnish layer on the top of the lignin derivatives layer gave significant difference in total coating agent deposition. In both the cases, lignin–glycerol and lignin–EG, total deposition was increased along with the number of attached lignin derivatives. Total deposition on the surface that was coated with lignin–glycerol after varnish coating was 2.77 ± 0.31 times than that of the samples that were only covered by lignin–glycerol. In case of lignin–EG, the amount was 2.03 ± 021. In other words, the amount of varnish polymer that was attached to lignin derivative coated surface compared to the derivative lignin itself for lignin–glycerol and lignin–EG was 1.77 ± 0.31 and 1.03 ± 021, respectively. It can be concluded that lignin derivatives helped the attachment of varnish polymer onto the wood surface. Surprising results were shown by lignin–glycerol that was able to retain varnish molecule with the mass almost twice of its own.

Partial substitution of the bottom layer coating agent was also observed beside the aforementioned fully substituted bottom layer. As much as 0.2–1.5 of samples ratio to the commercial product were mixed to meet 2 g of bottom layer mixture and then diluted into 10-ml commercial thinner. The resulted solution was then spread onto prepared 4×4-cm plywood by using 1-cm brush in the similar method to the aforementioned fully substituted bottom layer experiment. The weight difference was then noted as mass deposition on the surface and presented in mg/cm^2 and the results are shown in Figure 10.5.

In general, the conclusion for partial substitution of bottom layer was similar to the fully substituted one, lignin–glycerol derivatives gave larger deposit on the plywood surface compared to lignin–EG derivative. However, different trend was observed in these two experiments. The trend in lignin–glycerol derivative application in partial substitution gave the relatively similar amount of deposit on the plywood surface, that is, 6.5–9.5 mg/cm^2 when the ratio of lignin derivative is below 0.5. Significant increase of the deposition occurred when the ratio of glycerol–lignin derivative to commercial bottom layer is above 0.5, up to 10–13 mg/cm^2. On the other hand, lignin–EG derivatives have similar amount of deposition on the plywood surface in the whole range of mixture, that is, 4.5–6.5 mg/cm^2.

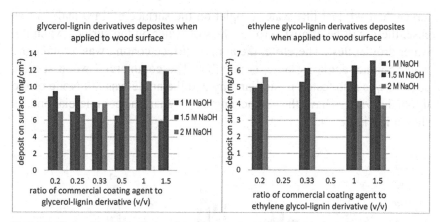

FIGURE 10.5 (See color insert.) Partial substitution effect of commercial coating agent with lignin derivatives to the amount of coating deposit on wood surface.

The performance of wood surface coating was analyzed by its ability to provide protection against water. This analysis was conducted by measuring the contact angle of water droplet on the wood-coated surface. Information technology significantly contributed to the technology that embedded in the Loop EXEMODE camera which able to enlarge image 3.5 times. Complete measurement required several softwares that were able to provide image modification. In some cases, converting the image into its negative form will help in the observation. Then, the software also required to provide virtual angle calculation as shown in Figure 10.6.

FIGURE 10.6 Example of photograph that was used for measuring water contact angle.

When the cohesion of water molecules is stronger than the adhesion of water molecules to the surface, the water droplet will have contact angle larger than 90, just like water droplet on the surface of Taro leaf. In this case, there will be almost zero residues of water on the surface when the water droplet flew. Therefore, the aim of surface coating was to make the water contact angle as large as possible. The commercial wood surface coating was also tested in similar way and the angle was 80. None of the lignin derivatives have the contact angle as high as the commercial one, as shown in Figure 10.7.

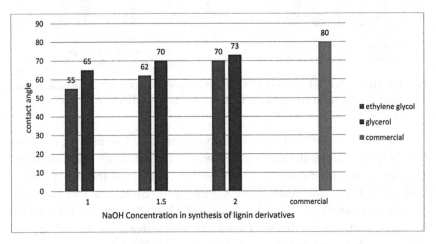

FIGURE 10.7   (See color insert.) The effect of NaOH concentration as catalyst in the synthesis to water contact angle when they were applied in double-layer coating, compared to commercial coating agent.

The results also showed that glycerol–lignin derivatives gave better protection from water due to its higher contact angle compared to ethylene glycol–lignin derivatives. This result also gave the strong suggestion that larger amount of deposition on the wood surface will give better water protection. This is very reasonable since the lignin derivative is a nonpolar compound. Lignin, in nature, is the protector of xylem so that water inside xylem will not leak out but flow through the stem to the highest leaf.

However, lignin derivative in this study had darker color. It can be observed in Figure 10.8. Thus, caution needs to be taken when the water protection is compromised by the esthetical preferences.

FIGURE 10.8 (See color insert.) Colors of the wood surface when glycerol–lignin derivative was applied. Increasing deposition and concentration will give darker color.

10.3 CHEMINFORMATICS IN LIGNIN DERIVATIVES PRODUCTION PATH ASSESSMENT

Lignin derivatives production path that observed in this article was embedded in te aforementioned bioethanol production process and can be seen in Figure 10.9. Main additional unit for this part was multieffect evaporator and recovery boiler. These units had already been familiar among pulp and paper producers since they are significantly increasing plant efficiency and reducing bad environmental impact. Even when these recovery units were underappreciated, in fact, over 1.3 billion t of weak black liquor are processed annually. After evaporation, about 200 million t of dry solid black liquor is burned in recovery boilers and recovering almost 50 million t of caustic soda as it produces 700 million t of high-pressure steam. According to these numbers, black liquor is the fifth most important fuel in the world, next to coal, oil, natural gas, and gasoline (Tran and Vakkilainnen, 2007). Lignin from palm oil EFB has a similar characteristic with softwood lignin which is about 5% higher in its heating value.

The computational simulation had been conducted in studying conceptual plant by converting typical Scandinavian kraft pulp mill to a biorefinery ethanol producing plant from cellulose. This simulation was conducted by using "future resource-adapted mill" which was designed within Swedish national research programme. The simulation software was using WinGEMS and analysis of the simulation results was conducted by Innventia (formerly STFI-Packforsk) (Fornel and Berntsson, 2012). This previous work clearly stated the importance of cheminformatics in enhancing chemical engineering analytical results.

Critical information that was required for this simulation was raw material composition. In this previous study, the raw material was assumed to be spruce with cellulose, hemicellulose, and lignin which were 40, 27, and 26%, respectively. By using this raw material, the obtained pulp was assumed to be 50%, ethanol conversion was 56% of theoretical yield or equal to 17% of the raw material. To perform energy, excess biorefinery in this simulation, investment was required to install latest lignin evaporation unit which was upgrading the 5.5 effect steam economy to 7. By this installation, the steam outlet can be used to operate condensing turbine with significant electricity yield. In capacity setting, 1800 dt/day of processed raw material or 390 kl of bioethanol, condensing turbine estimated electricity product was 5 MJ/l-bioethanol. This estimation was built because all lignin was burnt to produce steam. In other simulation where some part of lignin was separated as side products, estimated electricity product was 2.2 MJ/l-bioethanol and 8.2 t lignin/day. This production can be optimized by applying energy efficiency which reduces steam demand of overall process by 35–40%. By this steam saving design, less lignin demand for fuel in recovery boiler and it was estimated that 17.1 t lignin can be extracted (Fornel and Berntsson, 2012).

Purified lignin powder was prepared by using acid precipitation technology which is similar to the one that was assessed by (Levasseur et al., 2011). In this assessment study, the opportunities for using CO_2 in acidification were observed. Integration of bioethanol production plant with lignin and lignin derivatives production plant will increase the possibility of the CO_2 source by using ethanol fermentation as a side product. Although the CO_2 from fermentation as side product was available in abundance, it existed in low flow rate and pressure. Thus, if the CO_2 recycling loop was applied in this production process, then CO_2 from recovery boiler or lime kiln as suggested by Levasseur et al. was better option. From the existing observed mill which have raw material processing capacity 367 dt/day, CO_2 from recovery boiler was available with flow rate 14.7 kg/s and CO_2 fraction 22.3%w. CO_2 from lime kiln was available in less flow rate, 1.7 kg/s but with higher fraction 35.6%.

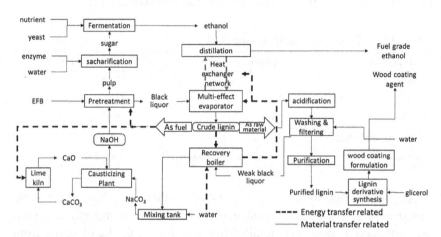

FIGURE 10.9   Block diagram for the integrated system of bioethanol production plant. The main raw material was palm oil EFB and the product was fuel grade ethanol and wood coating agent.

The main challenge in applying CO_2 recycling loop integration was high investment and energy demand compared to utilization of atmospheric air. A pair of monoethanolamide (MEA) absorber and regenerator column were required with their heat exchanger and pump auxiliaries. MEA regenerator column also has high energy demand. As much as 2.9–4.5 MJ/kg of CO_2 extract were required for the mentioned capacity. High fraction of CO_2 that is obtained from the regenerator column is surely a promising tool in increasing the effectiveness of acidification process.

Special attention was required in adopting all analytical results from both of aforementioned simulation. The difference in raw material composition from spruce that was used by Fornel and Berntsson with EFB that is discussed here can be observed in Table 10.1. This difference required several adjustments in assumption and calculation. Other important information was ethanol conversion. Fornel and Berntsson used only 57% of theoretical ethanol yield or similar with 17% yield calculated based on the overall raw material. Bioethanol pilot plant that was used as reference in this study was able to perform ethanol conversion up to 70% of theoretical yield (Sudiyani, 2011). Moreover, if the latest results in enhancing cellulase performance by surfactant addition were applied in the pilot plant, the conversion would reach 90% of theoretical yield or similar with 25% yield calculated based on the overall raw material (Bardant, 2017)

TABLE 10.1 Composition of Spruce and Palm Oil EFB.

Composition	Spruce	Palm oil EFB
Cellulose	40	33.25
Hemicellulose	27	35.3
Lignin	26	25.83
Extractive and other	7	5.62
Source	Fornell and Berntsson (2012)	Sudiyani (2011)

Lignin extraction that was conducted in this study had lower yield compared to previous studies since there is no evaporation applied. Lignin was obtained by only CO_2 acidification which gave 6.3 kg of lignin from 150 l black liquor or only 4.3%w. Application of LignoBoost Process would extract lignin up to 14.4%w of black liquor (Pertomani et al., 2011).

Cheminformatics approach that was used by the previous study in analyzing heat integration and followed by technoeconomic analysis was suitable to be applied in the existing bioethanol pilot plant that was observed in this study. Less C6 sugar in palm oil EFB can easily be compensated by applying latest research results. Significant investment was required to apply LignoBoost process to extract lignin as raw material for wood coating agent and fuel. Thus, multieffect evaporator with condensing turbine, recovery boiler, acidification, and purification was process units that required to be installed. Electricity required for these units can be covered by condensing turbine which was able to produce 2.2 MJ/l Bioethanol.

10.4 CONCLUSION AND FUTURE RESEARCH

Cheminformatics approach was suitable for preliminary analysis in planning plant intensification or product diversification. Several analyses from the previous study in transforming old pulp mill into biorefinery plant with ethanol production can be applied in a model of bioethanol production plant from palm oil EFB that was planned to be built in Indonesia. Information and research results from the existing bioethanol pilot plant gave significant support in developing cheminformatics preliminary analysis.

To build self-sufficient energy of integrated production plant which produces bioethanol, lignin, lignin derivatives, and wood coating agent, several additional units were required beside the bioethanol main process unit. Caustic regeneration plants were mainly required to perform caustic soda recycling into bioethanol pretreatment unit. This plant is a common part of any pulp mill that uses kraft pulping process, thus, all involved technology was well established with 80–90% of caustic recovery. Since no sulfur was involved in pretreatment pulping process, then thread of Burkeite ($2Na_2SO_4 \cdot Na_2CO_3$) formation in the production line which caused scale or formation of SO_3 emission in flue gas will be beyond concern.

Main input for caustic regeneration plant then the energy for lime kiln. This energy requirement can be fulfilled by installing multieffect evaporator and recovery boiler. Multieffect evaporator produced lignin that can be used as fuel for the lime kiln and raw material for lignin purification process. Lignin slurry with 60–65%w of solid that was obtained from multieffect evaporator can be used as fuel for heat generation in the evaporator itself, fuel for the lime kiln and fuel for pretreatment process. The weak liquor fuel potency can further be used in recovery boiler by direct burning. The obtained heat can be stored as steam which can be used further in heat exchanger network, including multieffect evaporator. The molten salt of $NaCO_3$ from the burning process was passed to the causticizing plant.

Product diversification unit could provide pure lignin, lignin derivatives, and wood coating agent as its product variations. Which product that feasible to be developed and at what capacity will be the main concern of future research. Demand of pure lignin on commercial scale is rare since most plant that uses lignin prefer to purify lignin by themselves. Thus, production of lignin derivatives and application of lignin derivatives as an active component in wood coating agent became a promising route for diversification. As shown in this article, production of lignin derivatives as an integral part of bioethanol production plant increase the overall energy efficiency which leads to the increasing feasibility of integration process. Research results that served in this study were a good start in analyzing the coating performance. Further research should be conducted in market acceptability and public aesthetic preference.

KEYWORDS

- empty fruit bunch
- cheminformatics
- automatic-computerized system
- lignin
- lingocellulose-based bioethanol

REFERENCES

1. Aldinucci, M.; Campa, S.; Danelutto, M.; Vanneschi, M. et al. (2008). Behavioral Skeletons in GCM: Autonomic Management of Grid Components, Italy: University of Pisa, p 56.
2. Barati, M.; Lotfi, S.; Rahmati, A. A Fault Tolerance Algorithm for Resource Discovery in Semantic Grid Computing using Task Agents. *J. Software Eng. Appl.* **2014**, *7*(4), 256–263. http://search.proquest.com/docview/1524713469?accountid=9765
3. Benedict, S. Performance issues and performance analysis tools for HPC cloud applications: A survey. Computing. Archives for Informatics and Numerical Computation **2013**, *95*(2), 89–108. http://search.proquest.com.ezproxy.ubd.edu.bn/docview/1283134915/DC779DBE04CE41DFPQ/13?accountid=9765
4. Bingxiang, G.; Lihua, J. Research on Grid Technology Applied in the China Financial Sector. *Future Communication, Comput. Control Manage.* Lecture Notes in Electrical Engineering, **2012**, 1–6. http://link.springer.com/chapter/10.1007/978-3-642-27314-8_1#page-1
5. Cope, J.; Iskra, K.; Kimpe, D.; Ross, R. Grids and HPC: Not as Different as You Might Think? I/O Forwarding in Grids. *State of the Art in Scientific and Parallel Computing.* 2010.http://www.mcs.anl.gov/research/projects/iofsl/pubs/para10-extabs.pdf
6. Dayyani, S.; Khayyambashi, M. R. A Comparative Study of Replication Techniques in Grid Computing Systems. *Int. J. Comput. Sci. Inf. Secur.* **2013**, *11*(9), 64–73. http://search.proquest.com/docview/1468454387?accountid=9765
7. Dongarra, J. J.; van der Steen, A. J. High-Performance Computing Systems: Status and Outlook. *Acta Numerica* **2012**, *21*, 379–474. http://search.proquest.com.ezproxy.ubd.edu.bn/docview/1002683861/CF6569B03AAD41E4PQ/2?accountid=9765
8. Eadline, D. *High Performance Computing For Dummies;* 2009. http://hpc.fs.uni-lj.si/sites/default/files/HPC_for_dummies.pdf
9. Eurich, M.; Calleja, P.; Boutellier, R. Business Models of High Performance Computing Centres in Higher Education in Europe. *J. Comput. Higher Edu.* **2013**, *25*(3), 166–181. http://search.proquest.com.ezproxy.ubd.edu.bn/docview/1448980286/CF31C07F1D304E14PQ/8?accountid=9765

10. Garg, S. K.; Sharma, B.; Calheiros, R. N.; Thulasiram, R. K.; Thulasiraman, P.; Buyya, R. Financial Application as a Software Service on Cloud. *Communications Comput. Inf. Sci. Contemp. Comput.* **2012**, *306*, 141–151.

11. Geist, A.; Reed, D. A. A Survey of High-Performance Computing Scaling Challenges. *Int. J. HPC Appl.*, **2015**, *3*. http://hpc.sagepub.com.ezproxy.ubd.edu.bn/content/early /2015/08/05/1094342015597083.full.pdf html

12. Goldsborough, R. Computing in the Cloud. *Tech Dir.* **2010**, *70*(5), 14. http://search. proquest.com/docview/819261026?accountid=9765

13. Guerrero, G. D.; Imbernón, B.; Pérez-Sánchez, H.; Sanz, F.; García, J.,M.; Cecilia, J. M. A Performance/Cost Evaluation for a GPU-Based Drug Discovery Application on Volunteer Computing. *BioMed Res. Int.* 2014. http://search.proquest.com.ezproxy. ubd.edu.bn/docview/1552819722/BAB0EEFD32ED4A77PQ/9?accountid=9765

14. IDC: The Changing Face of HPC. 2015. http://www.hpcwire.com/2015/07/16/idc-the-changing-face-of-hpc/ (accessed February 25, 2016).

15. Jacob, B.; Brown, M.; Fukui, K.; Trivedi, N. (2005) *Introduction to Grid Computing;* IBM Redbooks, **2005**, pp 3–6. https://www.redbooks.ibm.com/redbooks/pdfs/ sg246778.pdf

16. Joseph, J.; Fellenstein, C. *Grid Computing;* Upper Saddle River, NJ: Prentice Hall Professional Technical Reference, 2004.

17. Karthikeyan, P; Manjula, K. A. (2010). Business Applications of Grid Computing: a Review. *Natl. J. Advances in Compt. Manage.* **2010**, *1*(2). http://search.proquest. com/docview/1765404098?accountid=9765

18. Li, K.; Tsai, Y.; Tsai, C. Toward Development of Distance Learning Environment in the Grid. *Int. J. Distance Edu. Technol.* **2008**, *6*(3), 45–57. http://search.proquest. com/docview/201699518?accountid=9765

19. Liu, L.; Zhu, D. An Integrated e-Service Model for Electronic Medical Records. *Inf. Syst. eBus. Manage.* **2013**, *11*(1), 161–183. http://http://search.proquest.com.ezproxy. ubd.edu.bn/docview/1283523065/6CD5F6439BED4A8APQ/6?accountid=9765

20. Mustafee, N. Exploiting Grid Computing, Desktop Grids and Cloud Computing for e-Science. *Transforming Gov.: People, Process Policy* **2010**, *4*(4), 288–298. http:// search.proquest.com.ezproxy.ubd.edu.bn/docview/761437812/D40BAF0295584220 PQ/1?accountid=9765

21. Nicoletti, B. *Cloud Computing in Financial Services;* New York: Palgrave Macmillan, 2013.

22. Oesterle, F.; Ostermann, S.; Prodan, R.; Mayr, G. J. Experiences with Distributed Computing for Meteorological Applications: Grid Computing and Cloud Computing. *Geosci. Model Dev.* **2015**, *8*, 2067–2078. doi:10.5194/gmd-8-2067-2015

23. Ranilla, J.; Garzón, E.; Vigo-Aguiar, J. High Performance Computing: an Essential Tool for Science and Engineering Breakthroughs. *J. Supercomputing* **2014**, 511–513. doi:10.1007/s11227-014-1279-6

24. Rodero, I.; Parashar, M. Cross-layer Application-aware Power/Energy Management for Extreme Scale Science. Rutgers University, Rutgers Discovery Informatics Institute and NSF Cloud and Autonomic Computing Centre, 2012.

25. Sanchez, J. M. Global Behavior Modeling: a New Approach to Grid Autonomic Management. Ph.D. Thesis, University of Politecnic, Madrid, 2010, pp. 16, 65–70.

26. Selvi, S. Resource Management System for Computational Grid. *International J. Multidiscip. Approach Stud.* **2015**, *2*(4), 1–4.

27. Shawky, D. Scalable Approach to Failure Analysis of High-Performance Computing Systems. *ETRI J.* **2014**, *36*(6), 1023–1031. doi:10.4218/etrij.14.0113.1133

28. Shi, A.; Xia, Y.; Zhan, H. *Applying Cloud Computing in Financial Service Industry* [Abstract]. In 2010 International Conference on Intelligent Control and Information Processing, 2010.

29. Stanoevska-Slabeva, K.; Wozniak, T.; Ristol, S. *Grid and Cloud Computing: A Business Perspective on Technology and Applications;* Springer: Heidelberg, 2010.

30. Vecchiola, C.; Pandey, S.; Buyya, R. In *High-Performance Cloud Computing: A View of Scientific Applications* [Abstract], 10th International Symposium on Pervasive Systems, Algorithms, and Networks, 2009.

31. Vile, A.; Liddle, J. *TheSavvyGuideTo HPC, Grid, Data Grid, Virtualisation and Cloud Computing;* 2008; ISBN 978-0-9559907-0-0.

32. Zhang, Y. *Future Communication, Computing, Control and Management;* Springer Berlin: Berlin, 2014; Vol. 2.

CHAPTER 11

A SHIFTING PARADIGM OF A CHEMISTRY METHODS APPROACH: CHEMINFORMATICS

HERU SUSANTO[1,2,*], CHING KANG CHEN[3], TEUKU BEUNA BARDANT[1], and ARIEF AMIER RAHMAN SETIAWAN[1]

[1]*The Indonesian Institute of Sciences, Jakarta Timur, Indonesia*

[2]*Tunghai University, Taichung, Taiwan*

[3]*School of Business and Economics, Brunei*

Corresponding author. E-mail: heru.susanto@lipi.go.id

ABSTRACT

Cheminformatics is the application of computer technology and methods for the chemical-related field in molecular modeling and computational chemistry. It helps in the drug-making activity and in the discovery of new drugs, also opens up new area for research development and education. Information and Communication Technology (ICT) play a major and crucial part in cheminformatics, especially to store the tremendous amount of data relating to chemical compounds and other chemical information. ICT also enable easy accessibility of these data because they can be easily obtained. Here, without ICT, chemists would not be able to create new drugs easily resulting in a huge setback in the drug discovery process. Before the development of cheminformatics storage for chemical compounds, traditional documentation system using books and documents were played. Getting all this information had always been a problem in chemistry. Chemists should be able to get access to accurate data to their desks whenever it is needed. Therefore, from the beginning, chemists have been developing documentation systems on chemical compounds. An effective data mining helps to create and study new chemical objects in which it allows authentication and checking of physical and chemical characteristics among a large collection of compounds.

11.1   INTRODUCTION

As time passed and technology improved, it was discovered that it was more efficient to store information on a computer desktop than just on the desk, hence, the birth of *cheminformatics*, which is the combination of chemistry and informatics. Initially, this field of chemistry did not have a name until 1998, when the term cheminformatics was first coined by Frank K. Brown. His definition is the "...mixing of those information resources to transform data into information and information into knowledge for the intended purpose of making better decisions faster in the area of drug lead identification and optimization." Generally, cheminformatics refers to the application of computer technology to anything that is associated with chemistry. Before the development of computerized data storage for chemical compounds, we had the traditional documentation system using books and documents. One of the earliest encyclopedias on chemical compounds is Beilstein's *Handbuch der Organischen Chemie* or *Beilstein's Handbook of Organic Chemistry,* whose first edition was published in the 18th century, consisting of two volumes, with more than 2000 pages and registering 1500 compounds. This comprehensive encyclopedia of organic structures covers chemical literature from 1771 to date.[13] Information system concept is one of the integral parts of cheminformatics. In this chapter, we will look at the history of cheminformatics and its applications and usefulness in the field of chemistry and beyond.[1–3]

11.2   MAIN IDEAS

11.2.1   HISTORY OF CHEMINFORMATICS

An effective data mining system helps to create and study new chemical objects in which it allows authentication and checking of physical and chemical characteristics among a large collection of described compounds. Getting all this information had always been a problem in chemistry. Chemists should be able to get access to accurate data to their desks whenever it is needed. Therefore, from the beginning, chemists have been developing documentation systems on chemical compounds.[4-6] The earliest and oldest journal on chemistry is *Chemisches Zentralblatt*, which appeared as early as 1830; another example is as mentioned above, the *Beilstein's Handbuch der Organischen Chemie* (*Handbook of Organic Chemistry and Chemical Abstracts*) has been published since 1907. One more notable example is March's *Advanced Chemistry*. The problem is these data storage systems are studied in basic chemical handbooks.

Chemical information branches wanted to benefit from computers as it is much more effective and efficient to save information or data on the computer than just on the desk. Therefore, computations and chemical information are two important components that made up cheminformatics. Polanski (2009)[13] stated that Johann Gasteiger, a famous German chemist, supported this statement with the fact that in 1975, *Journal of Chemical Documentation* (a journal specializing in chemical information compiled by American Chemical Society, one of the most authoritative providers of chemistry-related information) changed its name to *Journal of Chemistry and Computer Science*. Similarly, we can use the same title to show recent developments in this field of chemistry since the journal had just changed the name to *Journal of Chemical Information and Modeling* in 2004. Thus, the journal's history briefly illustrates the scope of the discipline of cheminformatics.

11.2.2 USEFULNESS OF CHEMINFORMATICS AND ITS APPLICATIONS

The main purpose of cheminformatics is to preserve and allow access to tonnes of data and information that are related to chemistry, moreover integrating information needed on specific tasks or studies. Another purpose is to aid in the discovery of new drugs.

Possible use of information technology is to plan intelligently and to automate the processes associated with chemical synthesis of components of the treatment in a very exciting prospect for chemists and biochemists. One example of the most successful drug discoveries is penicillin. Penicillin is a group of β-lactam antibiotic used in the treatment of infectious diseases caused by bacteria, usually Gram-positive manifold. The way to discover and develop drugs is the result of chance, observation, and many intensive and slow chemical processes. Until some time ago, drug design was considered labor-intensive, and the test process always failed.

Nowadays, drug abuse is becoming a major issue around the world. Cheminformatics contribute to this issue by predicting or designing drugs. Immunoassays are commonly used to detect or screen testing of drug abuse on individuals through their urine or other body fluids. Immunoassays screening commonly point out toward classes of drugs such as amphetamines, cannabinoids, cocaine, methadone, and opiates.

Moreover, drugs identifications can be done using specific details by applying mass spectrometry methods such as liquid and gas chromatography. Chromatography is the process suggested that cheminformatics is just a new name for an old problem (Hann and Green 1999). While some current interest in cheminformatics can come as natural enthusiasm for new things, tof separating constituents of a solution by exploiting different bonding properties of different molecules.

In cheminformatics, there is a good relationship between chemistry and technology. The development of information technology has evolved considerably over time. The development of information technology developed in conjunction with a variety of disciplines and applied in various fields. Advances in information technology make the information accessible quickly and precisely. Information technology has changed the way we do science. Surrounded by a sea of data and phenomenal computing capacity, methodologies and approaches to scientific issues are developed into a better relationship between theory, experiment, and data analysis.

Consequently, cheminformatics is related to the application of computational methods to solve chemical problems, with special emphasis on manipulation chemical structural information. As mentioned previously, the term was introduced by Frank K. Brown in 1998 but there has been no universal agreement about the correct term for this field. Cheminformatics is also known as chemoinformatics, chemioinformatics, and chemical informatics. Many of the techniques used in cheminformatics are actually rather well established, be the result of years if not decades of research in academic, government, and industrial laboratories. The main reason for its inception can be traced to the need for dealing with large amounts of data generated by the new approach to drug discovery, such as high-throughput screening and combinatorial chemistry. An increase in computer power, especially for desktop engine, has provided the resources to handle this flood. Many other aspects of drug discovery make use of the techniques of cheminformatics, from design of new synthetic route by searching a database of reactions known through development of computational models such as *quantitative structure–activity relationship* (QSAR), which associates something that is observed through the biological activity of chemical structures through the use of molecular docking program to predict the three-dimensional (3-D) structure protein–ligand complexes and then choose from a set of compounds for screening. One common characteristic is that this method of cheminformatics must be applicable to a large number of molecules.

Cheminformatics, which is the combination of chemistry and informatics, is obviously linked to computer applications. However, not all chemical branches that relied on computers should automatically be included in the field. Although this term was introduced in the 1980s, it has a long history with its roots going back to more than 40 years. The principles of cheminformatics are used in chemical representation and search structure, QSARs, chemometrics, molecular modeling, and structural elucidation of computer-aided design and synthesis. Each area of chemistry from analytical chemistry to drug design can benefit from the methods of cheminformatics. This chapter will briefly discuss the accomplishments in chemistry that information system and technology had supported.

11.2.2.1 INFORMATION TECHNOLOGY IN THE FIELD OF BIOTECHNOLOGY

Information technology applications in the field of molecular science have spawned in the field of biotechnology. Biotechnology is a branch of science that studies the use of living organisms (bacteria, fungi, viruses, etc.) as well as products from living organisms (enzyme, alcohol) in the production process to produce goods and services. This study is increasingly important because the development has been encouraging and impactful in the field of medicine, pharmacy, environment, and others. These fields include the application of methods of mathematics, statistics, and informatics to answer biological problems, especially with the use of DNA and amino acid sequences as well as the information related to it. Insistence of the need to collect, store, and analyze biological data from a database of DNA, RNA, or protein acts as a spur to the development of bioinformatics. There are nine branches of biotechnology, and one of them is cheminformatics. Cheminformatics is one of biotechnology's disciplines that is a combination of chemical synthesis, biological filtering, and data mining used for drug discovery and development.

11.2.2.2   DRUG DISCOVERY

Cheminformatics has developed ever since the time of its establishment throughout the decades when computers and technologies had also developed. Researchers and scientists have been developing a way of assembling information and data within computers. These innovations on computers and technologies have the ability to store and obtain chemical information.

The meaning of these disciplines mentioned above is the identification of one of the most popular activities compared to various fields of study that may exist below this field. Drug discovery requires some experts to bring in technology from other fields, and a great time to be provocative and give a long overdue paradigm shift.

The discovery and development of drug design is the result of the agreement, observation, and chemical processes which were less intensive and slow. Until some time ago, considered drug design would always use a labor-intensive process and the test failed (the process of trial and error). The possibility of the use of information technology to plan intelligently and to automate processes associated with chemical synthesis components of the treatment is a very exciting prospect for chemists and biochemists. Award to produce a drug that can be marketed faster is huge, so the target is at the core of cheminformatics.

The scope of cheminformatics is very broad from an academic point of view. Examples of areas of interest include planning synthesis, reactions and structure retrieval, modeling, 3-D structure retrieval, computational chemistry, visualization tools, and utilities.

11.2.2.3   CHEMINFORMATICS IN EDUCATION

In today's society, technology plays an important role, not only for unlimited access for communicating purposes but also education. Cheminformatics opens up a new field of education, although there are countless educational areas that exist today. Cheminformatics has numerous and distinct applications and database. Cheminformatics can be learned and taught online. This could give the society easy access to learning the disciplines and methods of cheminformatics. Universities have been partnering up with companies that offer higher education such as Coursera, Udacity, and edX where world experts can connect up with society with an internet connection and a computer for the purpose of tutoring or other educational purposes.

The use of information system is important to science educators; it can help them to do their task by exploring the technology that already exists. It can also make science educations that use information system in their learning become a higher education. It is important to note that the insertion of information system and technology into education or other courses will make a difference in higher education of the public school. Information system and technology is not a vehicle for change, but it is simply a tool used by its user to do certain objectives.

11.2.2.4 HEALTH INFORMATION SYSTEM

Health information system manages information related to health. Everything related to the health is managed by the information system in order to make it easy to be used. It also can help any activities of the organizations or individuals to access some information about their health easily. Some examples of it are hospital's patient administration system and human resource management information system. Overall, it is used by the organizations to integrate effort to collect, process, report, and use health information and knowledge to impact the policy and decision-making, program action, individuals and public health outcomes, and research.

The main function of the system is to set up some guidelines to help health workers, health information managers, and administrators at all levels to focus on increasing the effectiveness of time, accuracy, and reliability of health data. Although the emphasis might seem to be on hospitals' medical records, these guidelines have been designed to address all areas of healthcare where data are collected and information is generated. Guidelines describe activities that must be taken into consideration when responding to the question of data quality in healthcare, regardless of the setting. Readers are guided to access and enhance the quality of the data generated in the environment in which they function, regardless of size, isolation, or sophistication. These guidelines are intended for government policy makers and healthcare administrators in the primary-, secondary-, and tertiary-level healthcare. There are doctors, nurses, other healthcare providers and health information managers as well. All these share a responsibility for documentation, implementation, development, and management of health information services.

11.2.2.5 REDUCTION OF RISKS IN DRUG DEVELOPMENT

Cheminformatics can help chemists and other scientists to produce and manage information. In silico analysis using cheminformatics techniques can actually reduce the risks in drug development. Techniques such as virtual screening, library design, and docking figures go into the analysis. Physical properties that may have an impact on whether a substance has the potential to be developed as a drug are often examined by one of the cheminformatics features which makes comparison among a large number of substances. Examples are cLogP, a measure of the amount of molecular obesity in the system. Sometimes, conclusions can be drawn about a set of associated properties, such as when Chris Lipinski, an experienced medicinal chemist, formulated the famous *rule of five* which evaluates that compounds such as drugs tend to have five hydrogen-bond donors or fewer, 10 maximum hydrogen-bond acceptors, the calculated logP should be less than or equal to 5, and having a molecular mass of up to 500 Da. The compound which showed greater values than these criteria tend to have poor absorption or permeation, meaning the drug is not orally active drug in humans.[7-10]

11.2.2.6 FORECASTING COMPOUND AGAINST NEUROLOGICAL DISEASES

Many compounds kept in databases have already been explored for multiple aims as a part of drug discovery programs. Excavating this information can provide experimental evidence useful for structuring pharmacophores to determine the main pharmacological groups of the compound.[15,16] Predictor model and DrugBank predictor model dataset were built using data collected from the ChEMBL database by means of a probabilistic method. The model can be used to forecast both the primary target and off-targets of a compound based on the circular fingerprint methodology, one of the technologies developed by cheminformatics method. The study of off-target connections is now known to be as important as to recognize both drug action and toxicology. These molecular structures are the drug targets in the treatment of neurological diseases such as Alzheimer's disease, obsessive disorders, Parkinson's disease, and depression. In future, developing these multi-targeted compounds with selection and chosen ranges of cross-reactivity can report disease in a more subtle and effective way and will be a key pharmacological concept.

11.2.2.7 OPTIMIZING ANTIBACTERIAL PEPTIDES WHEN COMBINED WITH GENETIC ALGORITHM

Research done by Department of Medicine, University of Columbia, had correctly identified additional activity of peptides with 94% accuracy among the top-ranked 50 peptides chosen from an in silico library of approximately 100,000 sequences by adopting genetic algorithm. These methods let a radically increased capability to recognize antimicrobial peptide candidates. A genetic algorithm is well suited for difficulties involving string-like data method for search-and-approximation problems. Implementing an iterative method whereby computational and experimental methodologies are used to find a new improved starting point for initiating the genetic algorithms, which is a more effective tactic and training the machine to learn algorithms using the new data by improving the ability to predict peptide activity. Based on the research, it has been reported that there are several peptides that are active against pathogens of clinical importance.

11.2.2.8 POTENTIAL OF CHEMINFORMATICS DATABASE IN FOOD SCIENCE

Computer databases nowadays have become important tools in biological sciences. Bioinformatics tools support the identification of chemical compounds such as peptides and proteins by mass spectrometry, which is the most reliable tool for such applications.[17] Cheminformatics databases are generally used in biological and medical sciences and play an increasingly significant role in modern science. An emphasis on the developing character of data mining and management techniques in animal breeding and food technology made computer databases the most extensive resource for finding and processing such information. One of the main goals of cheminformatics is to clarify life from the chemical outlook. The biological activity of chemical compounds thus falls into both bioinformatics and cheminformatics; for example, food scientists and researcher will need to access databases to find out information about the biological activity and behavior of several food components.[18,19]

11.2.2.9 INCREASED EDUCATIONAL OPPORTUNITIES

People sometimes have difficulties to obtain formal qualification but yet are interested to learn about the methods of cheminformatics due to limitations such as time, finance, and commitment. Nowadays, modern technology online learning by video and web conferencing as well as free online learning are the resources for cheminformatics. Online learning is another option to be considered in learning cheminformatics and enabling the wisdom impact to a wide range of practitioners and disciplines and in time will increase rapidly growing communities around the globe. Online learning also increases the number of expertise with capability in many techniques and applications in the field of chemistry. Online courses have been offered in universities together with the collaboration with expertise in chemistry.[12,20]

The expanding significance of health, science, data, and informatics is evidence of the critical importance of cheminformatics in future. Using computer and connection, people easily have accessibility to cheminformatics education. Live discussion also plays a major role in how to share idea and thought in cheminformatics education in the globally connected world we live in.

11.2.2.9.1 Application and Software

In this chapter, applications of cheminformatics will be briefly highlighted. The applications included are the most commonly used which contribute in the field of cheminformatics.

11.2.2.9.2 Screen Assistant 2

Screen Assistant 2 (SA 2) is an open-source Java software dedicated to the storage and analysis of small to grand-size chemical libraries. SA 2 contains information and data on molecules in a MySQL database. Structured Query Language (SQL) is the common language that is used for the purpose of adding, collecting, regulating, and operating the content in a database. It is much preferred by a whole lot because of its quick response and processing that satisfy a lot of users.[11,14]

11.2.2.9.3 Bioclipse

In cheminformatics, there is a need of applications which can help users with the extensible tools in the obtaining and calculating process of what cheminformatics has to offer. Bioclipse contains 2-D editing, 3-D visualization, converting files into various formats, and calculation of chemical properties. All these are combined into a user-friendly application, where preparing and editing are easy such as copying and pasting, dragging and dropping, and redoing and undoing process. Bioclipse is in the form of Java and is based on Eclipse Rich Client Platform. Bioclipse has the advantages on other systems as it can be used in other forms based on the field of cheminformatics.

11.2.2.9.4 Cinfony

Toolkits such as RDKit, CDK, and Open Babel function very similar to each other but the difference is that each supports various sets of data formats. Although they have complementary features, operating these toolkits on similar programs is challenging because they run on different languages, different chemical models, and have different application programming interfaces.

Cinfony, a Python module that introduces all three toolkits in an interface, makes it easier for users to integrate methods and outcomes from any of those three toolkits. Cinfony makes it easier to perform common tasks in cheminformatics, tasks that include calculating and reading.

11.2.2.9.5 KNIME-CDK: Workflow-Driven Cheminformatics

Konstanz Information Miner (KNIME): workflow-driven Cheminformatics is one of modern data analytics platforms, which is open-source library and enables sharing and exposing the commodity. One of the features available is plug-in feature, which is efficient and easy to use and better suited, whereby enabling researcher to automate the routine task and data analysis and also enabling building additional nodes and data analysis pipelines from defined components that work well when combined with the existing molecule presentation. KNIME allows you to execute complex statistics and data mining by using tools, such as clustering and machine learning and even plotting and chart tools on the data to examine trends and forecast possible results.

One of the standard roles includes data manipulation tools to manage data in tables, for example, joining, filtering, and partitioning as well as executing this molecule transformation according to common formats. Other tools to manage data use are substructure searching, signatures generating, and fingerprinting for the molecular properties. KNIME also uses target prediction tool which can predict the effects of existing drug in terms of their toxicity by giving suggestion on what molecular mechanism is observed behind the undesirable side effects and repurposing by exposing the new uses of the existing drugs.

11.2.2.9.6 Chemozart: Visualizer Platform and a Web-based 3-D Molecular Structure Editor

Chemozart provides the ability to create 3-D molecular structures. This web-based application tool is also used for viewing and editing of these molecular structures. As modern technology evolved, Chemozart which have flexible core technologies can be accessed easily through a UR. The platform is independent and compatible and has been deliberately created in a way that it is compatible with the latest devices, that is, mobile. This application also enables the process of teaching given that it works on mobile devices. It is beneficial to students as it is user-friendly and they can easily understand the concept of stereochemistry of molecules while constructing, drawing, and viewing 3-D structures, therefore used for educational purpose. With the help of this web-based platform, the user can simply create as well as modify or edit or just view the structures of the molecular compounds by just rearranging the position of atoms as simple as dragging them around or using keyboard, or now, using any touchscreens devices.

11.2.2.9.7 Open3DALIGN: Software Focuses on Unsupervised Ligand Alignment

One of the classical tasks of cheminformatics is unsupervised alignment of a structurally varied series of biologically active ligands which leads to various ligand-based drug design methodologies. The most important ligand-based drug design methods are pharmacophore elucidation and 3-D QSAR studies. Open3DALIGN together with its scriptable interface has the capability of carrying out both conformational searches and many unsupervised conformational alignments of 3-D molecular structures rigid body which makes automated cheminformatics workflows an ideal component of high throughput. Now, different algorithms have been applied to perform single- and multi-conformation superimpositions on one or more templates. Alignments which contain two operations feature matching and conformational search can be achieved by corresponding pharmacophores and heavy atoms or any combination of the two. Feature matching can be accomplished through field-based, pharmacophore-based, and atom-based methods approaches whether to find a same matching molecular interaction fields or searching a collection of pharmacophoric points or heavy atom pairs. Finding the best suited conformer for each ligand which may be mined from pre-built libraries are the strategy used in conformational search and can be achieved by following rigid alignment on the template and candidate ligands which may also be easily adaptable and aligned on the template. Regardless of the methods and approaches, great computational performance has been achieved through well-organized parallelization of the code features.

11.2.2.9.8 CDK-Taverna: Open Workflow Environment Solution for the Biosciences

Computational process and analysis of small molecule essential for both cheminformatics and structural bioinformatics, for example, in drug discovery application. CDK-Taverna 2.0 is the combination of unsimilar open-source projects whose goals are structuring a freely available open-source cheminformatics pipelining solution and becoming a progressively influential tool for the biosciences. CDK-Taverna 2.0 was effectively applied and tested and verified in academic and industrial environments with a sea of data of small molecules combined with workflows from bioinformatics statistics and images as well as analysis being made. CDK-Taverna supports the process of varied sets of biological data by constructing the complex-systems-biology-oriented workflows. In older days, insufficiencies like workarounds for iterative data reading were removed by sharing the previously accessible workflows developed by a lively community and that was available online, which enabled molecular scientists to quickly compute, process, and analyze molecular data as typically found in, for example, today's system biology scenarios. Graphical workflow editor is currently maintained and is being supported by design and manipulation of workflows. The features are considerably enhanced by the combinatorial chemistry-related reaction list. Implementing the identification of likely drug candidates are one of the additional functionalities for calculating score for a natural product similarity for small molecules. The CDK-Taverna project is recent and constantly updated and used in many ways by paralleled threads which are now enabled to carry out analysis of large sets of molecules and are faster in memory processing.

11.2.2.9.9 Open Drug Discovery Toolkit: ODDT

Drug discovery has become a significant element supplementing classical medicinal chemistry and high-throughput screening. This resulted in many computational chemistry methods which were developed to aid and learn capable drug candidates. ODDT is an open-source player in the drug discovery field which aims to fulfill the need for comprehensive and open-source drug discovery software because there has been enormous progress in the open cheminformatics field in both methods and software development. Unfortunately, there has only been little effort to combine them in only one package.

Structure-based methods are the most general and successful methods in drug discovery which are actively used to screen large small-molecule datasets, that is, online databanks or smaller sets (tailored combinatorial chemistry libraries). These methods are crucial for decision-making. Today, much effort is focused toward machine learning which is most valuable in clarifying both nonlinear and trivial correlations in data, respectively.

11.2.2.9.10 Indigo: Universal Cheminformatics Application Programming interfaces

Indigo is an open-source library which allows developers and chemists to solve many cheminformatics tasks. During the past years, due to a collection of more specific tools, we have made enormous development in the universal portable library. Tools suitable for scientists are popular programming languages as well as some GUI and command-line. Performance and important chemical features are the core of this C++ cheminformatics library. Among the chemical features of Indigo are its support to popular chemistry formats and *cis–trans* stereochemistry.

11.2.3 POSSIBLE FUTURES FOR CHEMINFORMATICS

Information system has changed the way we do things nowadays. The process of storing, finding the exact molecule by indexing, searching, retrieving information, and applying information about chemical compounds are made easy with modern technology. Cheminformatics uses computer and informational techniques which are applied to a range of problems in the field of chemistry, including chemical problems from chemical analysis and biochemistry to pharmacology and drug discovery. Cheminformatics contributions to decision-making are known to have helped certain scientists and chemist, for example, to extract right combination of density and structures from a database containing thousands of molecular structures data that are most likely to provide a specific function or healing effect. These are made available in the database to support research for better chemical decision-making by storing and integrating data in maintainable ways.

Improvement in technology assists cheminformatics in a way that now it has the potential to simulate protein complexes in solution, for example, pharmacophore analysis/visualization/pattern recognition, and also the most complex biological networks that were not possible with the use of pen and paper. Faster working capability of computer and high-speed network connections have developed the quality of algorithms, and eventually increase in the data sources. D. E. Shaw Group's recent publication shows that now millisecond simulations of drug docking using molecular dynamics are possible. This information can enhance knowledge worldwide. According to Glen Robert, there are altogether three areas that had to advance and develop in order to realize the possible growth of computational chemistry.

The first essential area is finding the most accurate and relevant data available. Collaboration among researchers, organizations that are involved in chemical research and relevant parties who are interested in discovery of new development in chemistry would make it possible that the relevant as well as credible and supported data and information are more open and available for everyone. These could be made possible with crowdsourcing and group of community sharing forum network as well as developing social system for sharing specific area of interest. Too much information could eventually become one of the main challenges to face. Today, in order to face the main challenge, language processing have been developed; even the most complex and complicated documents made by researcher can be processed by robots to extract and find the most useful information. These robots will assist us in finding the best suited data, then filter and organize them in a way they can cater our files of interest. Another possible solution will be developing networks that are self-organized and able to search and navigate the pool of data faster and relevant to our area of interest.

The second area of concern is how cheminformatics will be presented in computers—whether the highly professional researchers in chemistry are willing to share the information or they are bind by the confidentiality and secrecy of their work. These researchers tend to manipulate data by using complex language or symbols to present their research and chemistry concepts although the changes of name and description detail and easier term had been come up by three most influential in chemistry field, namely Berzelius, Archibald Scott-Couper and Frederick Beilstein almost 200 years ago. Now, the trend is slowly changing the name from complex to much simpler and the complex description become more easy to understand; the complicated detail attached to its function is made available in a simpler term yet there present is the accurate and relevant information with much more detailed information. The description of the information is not easily accessible but we can easily target the most relevant data that we are mostly curious in. More so, the symbols used in presenting molecules and structures are not easily made and understood. Proper indexing had also been made, although until now, it had not been fully incorporated and much improvement is still needed.

Third area is to replicate reality in the simulation. It is one of the main aims in computational chemistry. With modern technology, nowadays, petabyte computing is offered. It has the capacity to simulate the real biological systems within millisecond with hardware precisely designed for simulation. With vas development of hardware, the complexity of simulation would be made easier and user-friendly. There will be much improvement in accuracy as well. It is also dependent on the capability of the hardware design in future to cater the need for the development in simulation. When computing capacity is no longer limited in term of its capacity, extracting, filtering, and evaluating as well as recalculating could be done in more straightforward form and the complexity will no longer exist. IBM now is developing chips known as neural chips and with the emergence of cloud computing can improve further the area of cheminformatics. Super-high-speed connection, vast memory capability, and now, environmental virtualization had made it possible in future to change the current cheminformatics. With the modern technology where everything is done by technology it is possible that, according to Robert Glen, these machines will ask and think on their on to assist human in future.

Compounds will not be classified at molecular level anymore but rather in genetic and clinical effects term. Modern technology has been one of the major driving forces of the development of cheminformatics. It is hard to predict the future of cheminformatics but this chapter will cater for the possible future for cheminformatics and how the information system can support it.

11.2.3.1 FREE TEXT MACHINE LEARNING MODELS

Currently, it is difficult to process large information of data and extract information that is relevant to molecular discovery. In future, automation is expected to enable to extract relevant information and conclusions rather than relying on human curation.

The free text machine learning models are likely to enable machine learning process and statistical model generation and would combine multiple controlled vocabularies and ontologies, for example, chemical–biological and biological assays interactions in an appropriate manner.

11.2.3.2   INCREASE IN ENVIRONMENTAL CHEMINFORMATICS

Currently, the awareness for the society to study the relationship between environmental exposure and human health had increased from time to time. Nowadays, human beings are exposed to a wide range of environmental chemicals present in our air and water, integrally existing in our food, medicines, cosmetics, and many others.

Although these chemicals still exist at a low concentration of risk below the toxicological concern, this matter should be taken into serious consideration. This combination of chemical can potentially affect human health as fast as lightning. The integration of assessable data-monitored studies should cautiously take into account not just the concentration of chemicals but also the exposure time and life period through a collaborative framework. The expectation of chemical–chemical collaboration tools (here referring to their effects in living organisms, not to chemistry and physics) would be tremendously valued in risk assessment.

11.2.3.3   SIGNIFICANT DECISION-MAKING SUPPORT

In order to support decision makers to minimize risk of exposure, support and develop legislative and societal demands the forecast support systems could be used in evidence-based medicine to assess environmental threat and chemical warfare agents. The future system may assist with optimizing the selection of active medical ingredients during the process of formulating new drug mixtures.

The system will have the capability to handle combinations of drug mixtures of cheminformatics at the "fixed" and "unspecified" mixture level. These technologies of computer-assisted chemical production might become universal in our determination to lessen the cost of making drugs that are approved and to reduce the impact of chemical reagents on the environment.

11.2.3.4   DRUG DISCOVERY CONDUCTED BY AUTOMATED ROBOT

In future, robot scientists would perform procedures on the medicinal chemistry and toxicology works in search of additional proof to highlight the best likely chemicals and suggest these proofs to conduct cheminformatics experiments. These robots will identify safe chemicals that are supported by system chemical biology within the fundamental human context by using computer-aided molecular and synthesis proposal, for example, conducting biomolecular screening and toxicity experiments.

11.2.3.5 EXPANSION OF KNOWLEDGE AND OPEN-ENDED SCIENCE

Cheminformatics can assist the expansion of knowledge. The problem faced currently is the vast amount of molecular property, and sea of data often lead to difficulty in making choices and assumptions about the data essentially related to the quality and type of data input that is no longer relevant. Scientists with lack of experience are less likely to investigate multiple diverging options thoroughly while conducting their research. This situation is called epistemology and it is likely to occur when mining difficult data. This epistemological situation is a proof and is supported by anecdotal behavior, for example, most internet users prefer to take top 10 hits offered by search engines and do not bother to scroll down after 10 hits.

The development of integrative tools and approach based on knowledge within the Cytoscape framework will enable users to visualize molecular interaction networks by providing the basic features for data integration and analysis and share them through a public setting. The example of public setting is Crowdsourcing. Crowdsourcing can help knowledge sharing as it makes the collection of information easier. This tool will in future share and explore alternative hypotheses and multiple situations given the available limitations as well as building high confidence in predicting data and models, thus unlikely for the occurrence of epistemological situation.

11.2.3.6 INNOVATION AND THE IMPACT OF CHEMINFORMATICS ON SOCIETY

By combining cheminformatics with bioinformatics and other computational systems, it is expected to be beneficial to the developments in proteomics, metabolomics, and metagenomics, as well as other sciences. Now, in future, we will not need to estimate the bioactivity profile of a chemical at the molecular level but rather we will study biomedical information with the addition of inherent polymorphisms and clinical effects.

This development in proteomics, metabolomics, and metagenomics, as well as other sciences can be supported by developing tools for integrated chemical–biological data acquirement, filtering, and processing by taking into account relevant information related to collaborations between proteins and small molecules as well as possible metabolic alterations. These tools will be integrated into the virtual physiological human.

11.2.3.7   INCREASE IN COMPUTATIONAL SPEED

Graphics processing unit (GPU) technologies are likely to result in tremendous improvements over existing pharmacophore and fingerprint technologies when the cheminformatics software will be transferred on GPU platform and the ability to do research on highly accurate electronic densities will improve dramatically.

11.3   OPINION

Cheminformatics contributes to the community in many different ways. One of the ways is that it helps the discovery of new drugs and also to predict drugs. Cheminformatics helps researchers to experiment on chemical compounds and molecules through computers rather than to practically conduct the experiments. This can reduce the cost of the experiment.

Although cheminformatics have strong impacts on modern society and science, there are also many obstacles faced by eager and anxious users on learning cheminformatics. This could easily be a disadvantage for those keen for learning cheminformatics. The most common difficulty encountered is time limitation. Cheminformatics has various types of database and software that are installed in the computer. Amateurs will need to spend time to learn the way of how cheminformatics works, their database, and software. Thus, most people who learn cheminformatics have a shallow knowledge about it due to shortage of time.

Although experimenting with new drugs and discovering new drugs do not require high expenses, taking the course in learning cheminformatics requires high price of admission. Therefore, financial limitation is an important concern among learners in the field of cheminformatics.

There are many challenges in the implementation of cheminformatics. The world nowadays promotes the practice of going green and environment-friendly. It is crucial to search for the right chemicals which are the ones that have low toxicity and low environmental threat properties. Cheminformatics has been considered as highly reliable method for predicting the drugs for chemical combinations.

Other challenges faced by cheminformatics are that it should not be focused on chemistry alone. This is because chemicals also have influence and impacts on cellular functions which trickles a way to biological field.

11.4 CONCLUSION

Cheminformatics has a history almost as long as the computer itself. It is the application of computer technology and methods for the chemical related field is molecular modeling and computational chemistry. It is without a doubt that cheminformatics help humanity in many ways, especially in the drug-making activity and in the discovery of new drugs. Furthermore, it also opens up new area for research development and also increases opportunities for education.

Information system and technology play a major and crucial part in cheminformatics, especially to store the tremendous amount of data relating to chemical compounds and other chemical information. Information system and technology also enable easy accessibility of these data because they can be easily obtainable, whereas, in the older days, all these data and information were only kept in handbooks which made it hard to get the data when needed. Without information system and technology, chemists would not be able to create new drugs easily resulting in a huge setback in the drug discovery process. In the beginning, cheminformatics engineering has found particular application, especially in the pharmaceutical industry, but it is now beginning to penetrate into other areas of chemistry.

KEYWORDS

- **cheminformatics**
- **Information and Communication Technology**
- **drug discovery**
- **data mining**
- **chemical compound**
- **drug compound**

REFERENCES

1. Biesken, S.; Mein, T.; Wiswedel, B.; Figueiredo, L.; Berthold, M.; Steinback, C. KNIME-CDK: Workflow Driven Cheminformatics. *BMC Bioinf.* **2013**, *14*, 257.
2. Brown, F. Editorial Opinion: Chemoinformatics—A Ten Year Update. *Curr. Opin. Drug Discovery Dev.* **2005**, *8*(3), 296–230.
3. Churinov, A.; Savelyev, A.; Karulin, B.; Rybalkin, B.; Pavlov, D. Indigo: Universal Cheminformatics API. *J. Cheminf.* **2011**, *3*(Suppl. 1), P4.
4. Ekins, S.; Gupta, R.; Gilford, E.; Bunin, B.; Waller, C. Chemical Space: Missing Pieces in Cheminformatics. *Pharm. Res.* **2010**, *27*, 2035–2039.
5. Fjell, C.; Jenssen, H.; Cheung, W.; Hancock, R.; Cherkasov, A. Optimization of Antibacterial Peptides by Genetic Algorithms and Cheminformatics. *Chem. Biol. Drug Des.* **2010**, *77*, 48–56.
6. Fourches, D. Cheminformatics: At the Crossroad of Eras. *J. Cheminf.* **2014**, *8*, 16.
7. Glen, R. Computational Chemistry and Cheminformatics: An Essay on the Future. *J. Comput. Aided Mol. Des.* **2011**, *26*, 47–49.
8. Hann, M.; Green, R. Chemoinformatics—A New Name for an Old Problem? *Curr. Opin. Chem. Biol.* **1999**, *3*(4), 379–383.
9. Le Guilloux, V. Mining Collections of Compounds with Screening Assistant 2. *J. Cheminf.* **2012**, *4*, 20.
10. Micinski, J.; Minkieicz, P. Biological and Chemical Databases for Research into the Composition of Animal Source Foods. *Food Rev. Int.* **2013**, *29*, 321–351.
11. Nikolic, K.; Mavridis, L.; Oscar, M.; Ramsay, R.; Agbaba, D.; Massarelli, P.; Rossi, I.; Stark, H.; Contelles, J.; John, B.; Mitchell, O. Predicting Targets of Compounds Against Neurological Diseases Using Cheminformatic Methodology. *J. Comput. Aided Mol. Des.* **2014**, *29*, 183–198.
12. O'Boyle, N. M.; Hutchison, G. R. Cinfony—Combining Open Source Cheminformatics Toolkits Behind a Common Interface. *Chem. Cent. J.* **2008**, *2*, 24.
13. Oprea, T.; Taboureau, O.; Bologa, C. Of Possible Cheminformatics Futures. *J. Cheminf.* **2011**, *26*, 107–112.

14. Polanski, J. Cheminformatics. Elsevier B.V.: USA; **2009**, *4*, 14.
15. Sajadi, F.; Mohebifar, M. Chemozart: A Web Based 3D Molecular Structure Editor and Visualizer Platform. *J. Cheminf.* **2015**, *7*, 56.
16. Shiri, F.; Balle, T.; Tosco, P. Open3DAlign: An Open Source Software Aimed at Unsupervised Ligand Alignment. *J. Comput. Aided Mol. Des.* **2011**, *25*, 777–783.
17. Siedlecki, P.; Zielenkiewicz, P.; Wojcikowski, M. Open Drug Discovery Toolkit (ODDT): A New Open-Source Player in the Drug Discovery Field. *J. Cheminf.* **2015**, *7*, 26.
18. Spjuth, O. Bioclipse: An Open-Source Workbench for Chemo- and Bioinformatics. *BMC Bioinf.* **2007**, *8*, 59.
19. Truszkowski, A.; Jayaseelan, K.; Neumann, S.; Willighagen, E.; Zielesny, A.; Steinback, C. New Developments on the Cheminformatics Open Workflow Environment CDK-Taverna. *J. Cheminf.* **2011**, *3*, 54.
20. Wild, J. Grand Challenges for Cheminformatics. *J. Cheminf.* **2009**, *1*, 1.
21. Wild, J. Cheminformatics for the Masses: A Chance to Increase Educational Opportunities for the Next Generation of Cheminformaticians. *J. Cheminf.* **2013**, *5*, 32.

CHAPTER 12

INFORMATICS FOR SCIENCES: A NOVEL APPROACH

HERU SUSANTO[1,2,*], CHING KANG CHEN[3], and
TEUKU BEUNA BARDANT[1]

[1]The Indonesian Institute of Sciences, Jakarta, Indonesia

[2]Tunghai University, Taichung, Taiwan

[3]School of Business and Economics, Bandar Seri Begawan, Brunei

*Corresponding author. E-mail: heru.susanto@lipi.go.id

ABSTRACT

Information system (IS) primary purpose is to collect raw data and transform it into information that is beneficial for further processes. Main roles of IS are to support the competitive advantage, business decision-making, business processes, and operations. Each role deal with its own strategic level, tactical level, and operational level. In fact, it varies depending on the usage of the information as a driver of science activities. Decision support system is used for decision-making where the system will analyze and work on existing data which will project statistical prediction and data model. Transaction process systems allowed multiple transactions such as collecting, processing, storing, displaying, modify, or cancelling transaction at one time.

The difficulty to find the result of relationships between genetic variability, diseases, and treatment responses where it usually comeout as uncertain and inconsistent. Here, the system that may increase the size of the classical rule engines such as Fuzzy Arden Syntax or the probabilistic OWL reasoner Pronto is applied.

Interconnection between software and hardware during collecting and processing signal is well-detected by laboratory equipment, charged-couple devices, and spectrophotometers that are used to measure the amount of light reflected or absorbed from a sample object and other devices that can be connected to a computer through an analog to digital. Computer-aided algorithms used to examine the behavior of thousands of genes and creating a foundation of data for building integrated models of cellular processes. Experiments with various computational methods established in artificial intelligence, including knowledge-based and expert systems, qualitative simulation, and artificial neural networks. The outcomes have been obtained in finding IS and informatics has proved to have conspicuously great impacts on different areas of sciences and have find-out active genes in genomic sequences, assembling physical and genetic maps, and predicting protein structure.

12.1 INTRODUCTION

This chapter is exclusively concentrating on how information system (IS), technology, and science such as cheminformatics can be related to one another. This will also present in-depth understandings on how IS, technology can be used in conducting any scientific research with examples to be shown in the content section on the importance of studying bioinformatics, how data will be analyzed, and recent projects that are related to this topic.

12.2 INFORMATION SYSTEM

In an organization, they usually deal with thehuge amount of data. In order for the organization to organize and analyze the enormous data in a convenient and fastest way is by using IS. Many big companies such as eBay, Amazon, and Alibaba also use information system. Nowadays, the information is widely used in any organization. IS is a software that is used to collect, store, and process data. It is also used to spread information and feedback that will be provided to meet an objective. People often mistake data with information. Data are raw facts that are used for reference. Information, however, is a collection of data that can be used to answer the question as well as problem-solving (Paul Zanderbergen). IS primary purpose is to collect raw data and transform it into information that will be beneficial for decision-making.

The three main roles of an IS are to support the competitive advantage and decision making, business decision-making, and business processes and operations. For each role, it has its own level in which it operates. For support competitive advantage and decision-making, it works on a strategic level, while for support business decision-making it works on tactical level, and for support business process and operations it works on operational level. With regard to the type of IS, there is not only one type of IS. In fact, it varies depending on the usage of the information in an organization. Decision support system is one type of IS. Often, it is used for decision-making where the system will analyze and works on existing data which will project statistical prediction and data model. Transaction process systems allowed multiple transactions such as collecting, processing, storing, displaying, modify, or cancelling transaction at one time. Management IS where it mainly focuses on processing the data from transaction process in order to make it useful for business decisions according to specific problems. Office automation systems are concerned on office tasks as it will control the organization's flow of information. Executive ISs create conceptual information such as strategic and tactical decisions to satisfy the senior management. Expert system is said to be a set of computer programs that can imitate human expert.

In science, it is difficult to find the results of relationships between genetic variability, diseases, and treatment responses where it usually come out as uncertain and inconsistent. However, there are systems such as Fuzzy Arden Syntax or the probabilistic OWL reasoner Pronto that can be used to increase the size of the classical rule engines. Such systems are examples of decision support systems. It must be associated with specialists and international bodies so that it would affect the clinical practices efficiently. Based on these findings, the system we visualized must be directly interconnected with the IS of the hospital so that there is existing workflows when handling information gathered from electronic patient records and clinical laboratories.

12.3 INFORMATION TECHNOLOGY

Information technology (IT) is a computer technology that consists of hardware, software, computer network, users, and the internet. IT enables the organization to collect, make their work organize, and analyze data that helps them to achieve their objectives. IT personnel focus mainly on certain area, that is, business computer network, database management, information security, software development, and also sciences field. In bioscience, IT is generally used for automated data collection, statistical study of data, internet accessible shared databases, modeling and simulation, imaging and visualization of data and investigation, internet-based communication among researchers, and electronic dissemination of research results. For instance, IT being used for automated gene sequencers, which use robotics to process models and computers to manage, store, and retrieve data, have made potential the rapid sequencing of the human genome, which in turn has resulted in first time expansion of genomic databases. Shared internet accessible databases are important in paleontology, and models as well as databases are significantly used in population biology and ecology; and genomics are influencing many fields in biology. Furthermore, IT can be unique tools from the scientific tools for instance microscopes or physics accelerators, which are commonly used in the scientific process, such as data gathering. Additionally, IT supports in hypothesis formation that is the first stage to gather observations about the problem examine of biological study, research design, collection of data, data analysis, and communication of scientific result.

In other science field, IT also helps in analyzing subsurface creations, mapping, and modeling complex systems. For example, seismic data used to measure earthquakes were traditionally recorded on paper or film but today it recorded digitally, making it possible for the researchers to analyze the data swiftly. Furthermore, internet-connected allow many researchers to obtain and contribute data to solve challenges. In several areas of sciences, imaging and visualization become important because it can give clear modeling that helps the researchers understand biological systems such as tissues, organisms, and cells.

The role of IT toward software is that it helps the software to communicate with the hardware to input data, process data, and give the output. The communication between software and hardware can be defined when collecting and processing signal detected by laboratory equipment, for example, charged-couple devices, spectrophotometers that are used to measure the amount of light reflected or absorbed from a sample object and other devices that can be connected to a computer through an analog to digital. Computer-aided algorithms are being used to examine the behavior of thousands of genes at a time and are creating a foundation of data for building integrated models of cellular processes. Molecular biologists and computer scientists have experimented with various computational methods established in intellect artificial, including knowledge-based and expert systems, qualitative simulation, and artificial neural networks and other mechanical learning techniques These methods have been applied to problems in data analysis, the creation of databases with advanced retrieval capabilities, and modeling of biological systems. Practical outcomes have been obtained in finding active genes in genomic sequences, assembling physical and genetic maps, and predicting protein structure. IT has proved to have conspicuously great impacts on different areas of bioscience.

12.4 BIOINFORMATICS

Bioinformatics is the application of computer technology in managing any biological information and their management. This application is extensively being used to analyze the biological and genetic information based on living things, both in plants and animals. Essentially, it has been used in many applications such as studying human disease, managing on biological information, managing ISs for molecular biology management, and other related fields. In other terms such as in biologically, it is the conceptualizing the study of molecules, composed with informatics techniques, combination of computer science, applied math and statistics in the sense of physical chemistry that are worked together.

Previously, the use of classical methods for studying diseases is mostly observed at single factors, but at the present with technologies, helps in studying multiple factors at the same time, together with each and every possible thousand of variables. The application of bioinformatics methods also being used to undo the fundamental molecular biology of the disease and move toward personalized medicine in which, it is involved in understanding these diseases and transfer to new and more fitting treatment regimes.

The purposes of studying bioinformatics are that any biological data will be organized in a specified database, and this will help researchers to access or even enter any new related entries accordingly. Another, it is useful tool in developing the resources with the aid of data analysis, and lastly, this tool is able to develop, implement computational algorithms, and other software tools that help in understanding, interpreting, and analyzing any biological data that serve the humankind in a meaningful manner. Moreover, the computational tools are efficient on interpreting the results for biological research applied for protein, cell, and gene research, on discovering any new drugs development, herbicide-resistant crop combination.

The importance of bioinformatics application simply involved in melding biology with computer science, the use of genomics information in understanding human disease and the identifications of new molecular targets for drug discovery on a large scale. The use of bioinformatics are mostly in analyzing biological and genetic information which is associated with biomolecules on a large-scale, discipline in molecular biology areas from structural biology, genomics to gene expression studies applied to gene-based drug discovery, and development of genomic information resulting from the Human Genome Project. Experts and researchers furthermore practice bioinformatics application for studying the sequences of genomes which appear both in plants and animals in the field of agricultural studies, advancement in genomics technologies throughout the years from the 1950s until now. Genome sequencing apparently involves genetic databases for patients, and next-generation sequencing technology allows researchers to study complex genomics research, analyze genome-wide methylation, or DNA–protein interactions, and the study of microbial diversity in the environments and in humans.

Toward the end of this report, it will show how biological databases of commercialization on bioinformatics, biotechnology, and bioterrorism gives an impact to this new field of research and development for business purposes and current issues assist to this.

12.4.1 APPLICATIONS OF BIOINFORMATICS

12.4.1.1 GENOMICS, BIOMEDICINE, AND MICROBIOLOGY

According to Ma and Liu[5] genomics is a large-scale data acquisition, technological advancements that involve genome structures, evolution, and variations. Genomics origin can be traced as far back to the 19th century from the work of Gregor Mendel. However, in the middle of 19th century, the progress of IS and IT was not advanced as it is today. It is important to remember that genomics is an essential area of bioinformatics, as well as understanding its roles in the milestones of biological and molecular discovery. For instance, in Human Genome Project: an international scientific research project with goals to determine what makes up human DNA and its physical and functional characteristics, understanding heredity and diseases, and its role in pushing the innovation in genomic technologies, and many more. Another view on genomics is that the main concept of genome informatics was to analyze, process, and interpret all aspects of DNA in order to come up with a more defined and accurate information on biological structure and components of DNA.[1] All things considered, genomics is evolving duly because IT and IS keep on improving throughout the years. Owing to this, the world is progressing at a much faster pace, namely in biomedicine and microbiology, and the knowledge that it brought, had or are still being used to broaden our views on molecular mechanisms in the spreading, treating, curing, and preventing the development of diseases.

Almost every year, new drugs are being discovered or improved to better serve its purpose in curing, treating, and preventing health issues all around the world. Bioinformatics act as the main agent for its progression due to its vast collection of advanced tools for managing a large volume of data, as well as to help interpret, predict, and analyze clinical and preclinical data. As biological technologies progress, so does the data it produces and this often leads to a massive boom in database collection. Useful data could be proven to be useless without a proper system or tools to access it accurately and thus it is imperative to have computational tools that can search and integrate significant information. Development of bioinformatics resources has shown that it is essential in screening valuable data for effective drug solution in a profitable and timely manner. As an example, array comparative genomic hybridization method has been used globally for DNA analysis on normal and patho-logical clinical samples to check for the DNA copy number gain and loss across the chromosomes.[4] In addition, the current proteomic analysis that uses mass spectrometry-based technology is progressively used for iden-tifying molecular network targets and is responsible for many discoveries in profiling correlations in the pathogenesis of certain human illnesses. Through this analysis and method, integration of the connections between proteomic and metabolomics platforms can increase the dynamic and potency of the drug treatment solution.

Adverse drug reactions (ADRs) are often caused by all-purpose drugs that exist in today's market. Most people would prefer this type of drugs due to their economic status and often due to their state of living. Personalized medicine could provide the needed solution to these circumstances. Yan (2013) argued that through the development of pharmacogenomics and systems biology, personalized medicine could aid in the advancement of reductionism-based and disease-centered curative methods to systems-based, correlative and human-centered care. Developing further understanding between genotypes and phenotypes from analyzed data of genomic analysis through data integration methods help connects an efficient clinical and laboratory data flow. By implementing a translational informatics support into data mining techniques, knowledge discovery, and electronic health records, better diagnostic and treatment selection can bring a more suitable medication for the right people. A good translational bioinformatics will aid in establishing a powerful platform to connect various knowledge scopes for translating numerous biomedical data into predictive and preventive medicine.[10] Altogether, this will bring about an ideal personalized medicine that is less costly, reduce errors and risks, diminish ADRs, and overcome the therapeutic obstruction.

According to Wu,[9] cancer is one of the most prevalent and profound diseases that could occur at anytime and anywhere in the body. Its development is explained as an uncontrollable genetic mutation of cells in the body of organisms whereby it drastically affected the metabolism, loss in genes and promotion of invasive tumor growth, metastases and angiogenesis.[7] Multiple factors such as the period, severities, drug resistance, cell origins, locations, and affectability can be the causes for poor diagnoses and therapies result for the cancer patient. However, over the years, the results from advanced and accurate clinical bioinformatics and uses of new systems clinical medicine have helped improve the results of cancer treatment and diagnosis all around the world. Adoptive immunotherapy (gene therapy) is commonly used for cancer treatment, which uses the technology of genetic modification whereby T cells with antitumor antigen receptors (TCR) or chimeric antigen receptor (CAR) are used, duly because they can target antigens expressed on tumor cells.[6] In recent times, the use of Semantic web technology enable better understanding of high throughput clinical data and establish quantitative semantic models gathered from Corvus (data warehouse providing systematic interface to numerous forms of Omics data) rooted from systematic biological knowledge and by application of SPARQL endpoint.[2] In addition, application of new biomarkers strategies in cancer bioinformatics has become more popular in monitoring the progress of the disease and its response to therapy. It is expected to coordinate with clinical informatics which includes patient inputs such as complaints, history, therapies, symptoms and signs, medical examinations, biochemical analysis, imaging profiles, and other valid inputs. All in all, the expected result would provide more accurate interpretable signatures and therefore helps in better diagnosis and cancer solutions for specific patients.

The world's population is continuously booming and is expected to reach 11 billion by the year 2100 and this brings about new challenges in managing disease outbreaks. A rise of infectious diseases by new viruses and drug-resistant bacteria are the tendencies of disease outbreaks. Over the last two decades alone, new virus strains have kept the world in constant fear of deadly outbreaks threats: swine flu pandemic, severe acute respiratory syndrome, human immunodeficiency virus (HIV), and acquired immune deficiency syndrome (AIDS), malaria, tuberculosis, and more recently Zika virus. The steps in battling more outbreaks have been initiated throughout the globe but more importantly, the sharing of knowledge on how it occurred and the proper exploitation methods to discover the viruses' weaknesses has proven to be a better front in battling outbreaks. Next-generation genome sequencing has aided the advancement of biotechnologies and tools by providing new insight into viral distinctiveness, allows in-depth sampling and provides bigger capacity for automation, and thus providing new data interpretation on what could be done or changed to the characterization of viral quasispecies.[3] By Kijak et al.[3] their bioinformatics package named Nautilus, which runs on several operating systems, represents new sets of tools to support better data analysis to facilitate the application of next-generation sequencing and allowing better insight on HIV genome characterizations throughout the population and its evolution. Nevertheless, the rapid occurrences of antimicrobial resistance in microorganisms not only is diminished simply by continuous biological studies but also through a global understanding in public members to always keep a hygienic environment and in practices as well as awareness, wherever and whenever they are. This method is the main preventative method for generally, all kinds of microbial infections, duly because producing a constantly evolving medicine and treatment to fight against rapidly evolving antimicrobial resistance illnesses will take higher health care expenditures and is time-consuming for all sides. The risks of death from resistant microorganisms are much higher than that of the same nonresistant microorganisms.

12.4.1.2   BIOLOGICAL DATABASE

The database is seen as major tool for storing biological data for public use. Relational database concept of computer science and information retrieval concept of digital libraries is implemented to fully interpret biological database. Gene sequence, attributes, textual descriptions, and ontology classification are stored in the biological database. The data mentions are categorized as semi-structured data that later can be displayed in tables form, key delimited record, and XML structure. The common method of cross-referencing is often used by database accession number.

A biological database can be defined as the collection of biological data collected during live experiment and computation operation and analysis. Biological data should be organized properly to enable easy data operations such as manipulation, deletion, and calculation. The aim of the database must follow two principles which are accessible and can be used in both single and multiuser system environment.

Databases in common can be grouped into primary, secondary, and composite databases. Primary databases include data that is gathered during experiment such as nucleotide sequences and three-dimensional structure. It is often called as archival databases. The data gathered are resulted from experiment form researcher all around the world. GenBank, DNA database of Japan (DDBJ), SWISS-PROT. Secondary databases. Secondary databases are derived from analysis of the first-hand (primary) data such as sequences and secondary structure. The results of secondary databases are usually in the form of conserved sequences, signature sequences, and active site residues of the protein. The curated database is another term for secondary databases. Some of the databases were created and hosted by the researcher themselves at their own laboratories such as SCOP, CATH, and PROSITE which was developed in Cambridge University, University College of London, and Stanford University, accordingly.

The first databases were created in 1956 which was after the insulin protein sequence available which was the first protein to be sequenced. The content of the insulin sequence includes just 51 residues which characterized the sequence. After insulin protein, the first nucleic acid sequence of Yeast tRNA with 77 bases was discovered 4 year after which was in 1960. Three-dimensional structure and the creation of the first protein structure database with only 10 entries were studied in 1972. Currently, the Protein Data Bank (PDB) has grown to store more than 10,000 entries.

Since protein sequences databases were maintained in individual laboratories, SWISS-PROT protein sequence started to exist which was categorized as consolidated format database which begun in 1986. Database functionality was also expanded along with the capabilities of handling data, sophisticated queries facilities, and bioinformatics analysis function which were also implemented in modern databases.

Similar to general databases, the biological database can also be categorized into two groups, that is, sequence structure databases and pathway databases. Sequence structure databases are mainly focused on nucleic acid sequence and protein sequence where pathway databases will only focused on protein.

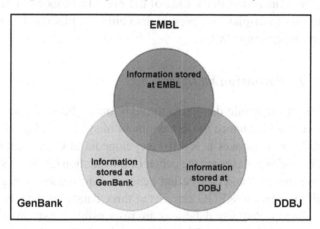

FIGURE 12.1 Information stored at GenBank, EMBL and DDBJ is shared with each database daily.

12.4.1.2.1 Sequence Databases

Sequence databases are categorized as the most frequently used databases and some of the databases in sequence databases are marked as the best biological databases. GenBanks is one of the examples of widely used sequence databases. GenBanks focuses on DNA and protein sequence.

GenBanks are classified as one of the most widely used sequence biological databases. The name refers to the DNA sequence databases of National Center for Biotechnology Information. GenBanks data are mainly made up of sequences submitted by individual laboratories and data interchange from international nucleotide sequence databases, European Molecular Biology laboratory, and DDBJ.[8]

12.4.1.2.2   Structure Databases

To completely understand the protein function, knowledge of protein structure and molecular interaction and mechanism is must. The PDB is the worldwide repository of experiments which classified the protein structure, nucleic acids, and complex assemblies, including drug-target complexes. The PDB was created in Brookhaven National Laboratories during 1971. It mainly contained information on the molecular structure of macromolecules obtained from X-ray crystallography and NMR method. Currently, Research Collaboratory for Structural Bioinformatics plays huge role in maintaining PDB. One of the main features of PDB is that it allows the user to display and present data either in plain text or through a molecular viewer using JMOL.

12.4.1.2.3   Pathway Databases

The growth of metabolic databases through metabolic study pathway will fulfill the need and enhance the development of system biology. One of the popular pathway databases is Kyoto Encyclopedia of Genes and Genomes (KEGG). KEGG databases are the center of information toward system analysis of gene function and connecting genomic information to higher order functional information. KEGG consists of three databases namely pathway, genes, and ligand. Pathway databases are responsible for storing the higher order functional information. These information include the computerized knowledge on molecular interaction networks. These data are often encoded by coupled genes on the chromosome which is crucial for predicting gene functions. The genes databases consist of the collection of genes catalog and sequence of genes and protein produced by the Genome Project. The third databases ligand stored information regarding the chemical compound and chemical reaction which is important to cellular processes.

12.4.1.2.4   The Commercialization of Bioinformatics

The success of development in combining both computer technology and biomedicine has helped scientist to become more efficient and productive with the ability to predict the upcoming trend of biosolution with the help of bioinformatics. With the advancement of bioinformatics technology widely being exploited, the opportunity of commercialization is unquestionable. However, many investors were reluctant to invest in bioinformatics sector due to its history of invest during the late 1990s which resulted in high loss. Dispute the loss being made, on 2002, Philip Green, biologist in University of Washington wanted to decipher human genome with more accurate reading of DNA letter. The celera-made machine was the only tools he used which were supplied by Applied Biosystems. Owing to the lack of functionality of this software, he then designed his own software to cater the needs of his project. As a result, Green's innovation is categorized as the industry standard and its source code is available without the need of paying.

12.4.1.2.5 Issues of Commercialization

The advancement of technology played a huge role in the success of bioinformatics where computer tools are being used for managing biological information and computational biology that is used to identify the molecular component of living things. The involvement of computer technology in the biological area also introduced the studies of principle and operation of data manipulation and data analysis of biomolecules, structure or composition, or various materials such as nucleic acids and product genes such as proteins. Research and data gathering on the biological field involved lab experiment where mathematical operations and computation are used to obtain meaningful information from meaningless data especially in genomics. Computers have been used as the backbone in bioinformatics and eventually becomes one of the major tools in storing biological data compared with the existing dataset which provide important useful input for the computer-user researcher that biologist usually gathers during their hands-on laboratory experiment saves a lot of time and resources. Cost of the laboratory has been cut down since the use of computer and software where most of the operation and experiments are done through computer with the help of special purpose software.

The popularity of using open-source software increased over the year which has affected bioinformatics companies. Open-source software can be defined as computer software which is freely available to the public. In general, it is free software which can be freely copied, distributed, modified, and manufactured. Characteristics of open source are: there should be no discrimination to people, groups or endeavor, the license distribution should be costless and general to every product and the license must be restriction free of other software and technology neutral. The Linux system is one of the great examples of open-source software with a major success. Bioinformatics firm finds it hard to gain profit due to this open-source software movement.

12.4.1.3   GENOMICS SEQUENCING IN AGRICULTURAL STUDIES

Another application of bioinformatics scientist has taken advantage of agriculture field in which the sequences of the plants and animals genomes offer great advantages for the agricultural. Moreover, this can be utilized for the genes within these genomes and their purpose is to produce healthier plants. Tools of bioinformatics play a significant role in providing the information about the genes which occurred in the genome of these species. These appliances have also made it possible to predict the purpose of various genes and factors affecting these genes. The information offered about the genes by the tools makes the scientists to produce an enhanced species of plants which have drought, herbicide, and pesticide resistance in them. In other word, bioinformatics plays a significant role which allows agriculture to enhance the food (plant) nutrient and improve crops that are capable to manage poor soil growing conditions and poor weather.

The relative genetics of the plant genomes has proved that the structure of their genes has continued to be more well-preserved in excess of evolutionary time than was previously thought. These results propose that evidence achieved from the prototypical crop schemes can be recycled to recommend progresses to other food crops. *Arabidopsis thaliana* and *Oryza sativa* (rice) are samples of comprehensive vegetation genomes.

Bacillus thuringiensis genes that can manage few of important nuisances have been effectively relocated to maize, potatoes, and cotton. This innovative facility of the vegetation to battle insect attack defines that the total of insecticides being utilized can be lessened and, therefore, the nutritional quality of the crops is greater than before.

Similarly, scientists have just flourished in transmitting protein sequence into rice with bioinformatics to upsurge levels of iron, vitamin A, and other micronutrients. Scientists have injected a gene from yeast into the tomato, and the outcome is a plant whose fruit got extended on the vine and has a lengthier lifespan.

Moreover, with technology, it supports to cultivate in subordinate soils and sturdy drought. The advancement has been completed in agricultural cereal diversities that have a greater leniency for soil alkalinity, iron toxicities, and free aluminium. These diversities will let food production to be resilient in poorer soil parts, consequently adding new land to the global production base. Furthermore, research is in progress to harvest crop diversities adept of enduring concentrated water conditions.

12.4.2   EVALUATION/OPINION

12.4.2.1   BIOTERRORISM

One type of threat that is possible due to the invention of bioinformatics is bioterrorism. Bioterrorism refers to terrorism involving the intentional release or dissemination of biological agents. These agents are bacteria, viruses, or toxins, and may be naturally occurring or human-modified, may result in the illness and even death of people, animals, and plants. These substances and agents are common in our natural environment; however, the properties of these agents can be altered to allow it to be more resistant to medicines and antibodies.

The main source of bioterrorism agents and substances disperse through air, water supply, and food supply. The terrorist may use these biological substances due to its detection difficulties in which the symptoms can only appear after several days or hours. Bioinformatics is covering to open access which can be easily accessed by anyone. These enable people with the bad intention to access biological data structure and analysis can be easily learned and altered to create a new virus. Owing to these issues, there will be a contradicting view on to making bioinformatics freely available and strictly to only specific intention for it not to fall to the wrong hand. Bioterrorism agent falls into three categories category A, category B, and category C. Category A include high-priority agent such as organisms or toxins which can cause harm and can be easily spread from person to person. It also may cause death, panic, and social destruction which require special action for public health. Category B is the second highest priority because it is fairly easy to disperse with the minimal cause of death. Category C is the highest priority agent where it can be used to engineer for mass spreading. These types of agents can be easily available, constructed, and dispersed without any restriction.

The effect of this category can result in the major impact of morbidity, mortality, and health. The use of bioterrorism and biological warfare long existed during 600 B.C. where the spread of diseases from animal and contaminating enemies water supplies became one of their enemies strategies. These strategies have continued and also being practiced during European wars, American Civil wars, and even now. During middle age, the infectious patient became very valuable which is perceived as a weapon. Siege of Caffa is one of the examples where the Tartar forces were infected with epidemic of the plague which then being spread to the city. This resulted in plague outbreak which caused the Genoese forces to retreat. The epidemic plague continued to spread all over Europe, the near East, and North Africa in the 14th century. This event is considered as the worst cases involving public health.

Another example of the use of the biological weapon was during March 1995 where the sarin gases were used to attack Tokyo subway system by the Aum Shinrikyo. Before this event, the group had attempted three attacks which were unsuccessfully executed with the use of anthrax and botulinum toxin. The members also tried to gain Ebola viruses in Zaire in the year 1992. However, the project had been uncovered by the Japanese forces. Unfortunately, the evidence caught was insufficient to be made public. Until now, the biological weapon project created by Aun Shinrikyo remained a mystery.

With the advancement of technology and biological breakthrough, it is quite hard to detect any bioterrorism activities. The outbreak of new virus and diseases has increased over a decade with the return of viruses which were long gone coming back to life. The cause and source of this virus remained uncertain whether from natural sources or from bioterrorist. To classify between those two are low in possibilities with the need for deep examination, analysis and taking into account of multiple perspectives such as current environment, original, and current structure of molecules.

12.4.2.2   ADVANTAGES AND DISADVANTAGES

Bioinformatics make the information accessible and shared in comparison to traditional biological records where the developing tools make it easy to send, receive, and share the information. For example, electronic medical records (EMR) reduce the opportunities of error that are caused by obstruction and other researcher's conflicts during the manual data entry process after data collection on paper. Besides that, it also supports to eliminate the manual task of extracting data from charts or filling out specified data sheets. The data stored can be obtained directly from the EMR. By referring to EMR, the researchers did not need to examine or observe the task again. Bioinformatics has grown rapidly, and divided into subdisciplines such as in chemistry named as cheminformatics and also neuroinformatics which is related to gathering data across all scales and neuroscience level to understand the complex function of brain and work toward treatments for brain-related illness and immunoinformatics, it uses informatics techniques to study molecules of the immune system.

Usually, the organization such as Antigen Discovery, Rasa Life Science Informatics, and LabCentrix that uses bioinformatics is storing huge amount of data or in other words, big data which means it contains both structured and unstructured data that is hard to process, if the organization uses traditional database and software techniques as the data are huge. Examples of big data are patient information, types of disease, and DNA. Hence, by applying bioinformatics this can be done easily as it can store huge amount of data.

From the definition of bioinformatics, it is said that bioinformatics uses computer technology. Since bioinformatics involved technology, thus, it can be a threat to the organization using it, as people nowadays know how to create computer threats and use it for the crime. Hence, bioinformatics can simply be hacked by computer hackers. Hackers are usually people who know programming really well and use the capability to break into someone's computer. There are different types of hacker such as white hat, black hat, grey hat, script kiddie, hacktivist, and phreaker. For bioinformatics, the organization has to be cautious on black hat hacker, that is, identity theft. Identity theft is when a person steals someone's information (such as name or date of birth) without their consent and takes advantage of it in order to get goods or services. Since bioinformatics contains a lot of confidential information, it is important for the organization to secure that information cautiously. One way to overcome this problem would be encrypting all sensitive files where each file will be secured with a password or a key and only people who know it can decrypt and get access to the file.

Bioinformatics interface seems to be unattractive, the designing of the tools have to be user-friendly where people can easily understand it and cope with it. To have a user-friendly interface, it can make the researchers analyze the information more efficiently and effectively to convert the information into knowledge. Other than that, the users must be properly trained to use bioinformatics tools to prevent difficulties. This would be a tough situation for developing regions as training and knowledge requires time and financial support. For example, in Ghana, they suffer months, and sometimes, years of drought, in which agriculture is almost impossible to succeed. Although bioinformatics might help them to grow plants, they may not be familiar with the system; hence, the knowledge about the technology is necessary.

Here, bioinformatics is important, yet the biggest challenge look out the molecular biology society nowadays is to make sense of the wealth of data that has been produced by the genome sequencing projects. With the advent of new tools and databases in molecular biology, researchers are now able to carry out research not only at genome level but also at proteome, transcriptome, and metabolome levels. Therefore, incisive computer tools must be improved to accept the extraction of meaningful biological information.

12.5 CONCLUSION

In conclusion, the applications of bioinformatics are widely used, from single traditional methods in handling genomic studies to more advanced methodologies toward the improvements in biological studies. Understanding that the vast collection of bioinformatics tools and technologies that exist to serve different purposes is also vital to the advancement in IT and IS and vice versa. Bioinformatics help in improving the ways of biological research and provide innovative ways for instance, the use of databases to store any biological information together with more advanced software tools and computational algorithm. These databases are mostly related to microorganisms that are being stored in the computer memory. In business studies, the research on scientific fields gives opportunities for businesses to broaden their opportunities in making profits. This instance the invention of new medicines, drug developments, and so forth.

In the world of biomedicine and microbiology, IT and IS of bioinformatics have provided numerous resources and tools into new drug discovery, personalized and preventive medicine. Moreover, the successes in cancer treatment through gene therapy are also gradually increasing, owing to the next-generation sequencing techniques and technologies. Not only for cancer, but also for other known diseases such as HIV and AIDS.

On the other hand, databases are also being widely used in bioinformatics application. The relational database concept of computer science and information retrieval concept is implemented to fully interpreted biological database. The aims of the database are basically accessible and can be used in both single and multiuser system environment. There are several types of databases provided for each and different purposes such as for protein development, human brain studies, diseases, drug discoveries, and development and also in agricultural studies such in both plants and animals. These data can be seen as highly accessible and reliable for other research and scientists. Not only that, it has data integration and even they can add new findings or even update as it is easy to use. Scientists and researchers are also being trained in handling databases.

Meanwhile, with bioinformatics, vegetations can successfully grow for the agricultural community. The tools used can help the plants to withstand drought season or poor weather condition, enhance the nutrition quality, resilient to pest, and poor soil environment.

Finally, even though the bioinformatics applications are widely being used for more advanced technological developments, there are pros and cons not only in biologically manner but also in business manner. As stated, these are being handled properly and their usage is being improved throughout the years.

KEYWORDS

- artificial intelligence
- knowledge-based
- expert systems
- decision support system
- computer-aided algorithms

REFERENCES

1. Chen, R. On Bioinformatic Resources. *Genomics Proteomics Bioinf.* **2015**, *13*(1), 1–3. http://dx.doi.org/10.1016/j.gpb.2015.02.002
2. Holford, M.; McCusker, J.; Cheung, K.; Krauthammer, M. A Semantic Web Framework to Integrate Cancer Omics Data with Biological Knowledge. *BMC Bioinf.* **2011**, *13*(Suppl 1), S10. http://dx.doi.org/10.1186/1471-2105-13-s1-s10
3. Kijak, G.; Pham, P.; Sanders-Buell, E.; Harbolick, E.; Eller, L.; Robb, M.; et al. Nautilus: A Bioinformatics Package for the Analysis of HIV Type 1 Targeted Deep Sequencing Data. *AIDS Res. Hum. Retroviruses* **2013**, *29*(10), 1361–1364. http://dx.doi.org/10.1089/aid.2013.0175
4. Leung, E.; Cao, Z.; Jiang, Z.; Zhou, H.; Liu, L. Network-Based Drug Discovery by Integrating Systems Biology and Computational Technologies. *Briefings Bioinf.* **2012**, *14*(4), 491–505. http://dx.doi.org/10.1093/bib/bbs043
5. Ma, D.; Liu, F. Genome Editing and its Applications in Model Organisms. *Genomics Proteomics Bioinf.* **2015**, *13*(6), 336–344. http://dx.doi.org/10.1016/j.gpb.2015.12.001

6. Morgan, R.; Chinnasamy, N.; Abate-Daga, D.; Gros, A.; Robbins, P.; Zheng, Z.; et al. Cancer Regression and Neurological Toxicity Following Anti-MAGE-A3 TCR Gene Therapy. *J. Immunother.* **2013**, *36*(2), 133–151. http://dx.doi.org/10.1097/cji.0b013e 3182829903

7. Mount, D. Using Bioinformatics and Genome Analysis for New Therapeutic Interventions. *Mol. Cancer Ther.* **2005**, *4*(10), 1636–1643. http://dx.doi.org/10.1158/ 1535–7163.mct-05–0150

8. Priyadarshi, M. B. Sequence Databases. Retrieved from http://www.biotecharticles. com/Bioinformatics-Article/Bioinformatics-Sequence-Databases-3278.html (Accessed Nov 10, 2014).

9. Wu, D.; Rice, C.; Wang, X. Cancer Bioinformatics: A New Approach to Systems Clinical Medicine. *BMC Bioinf.* **2012**, *13*(1), 71. http://dx.doi.org/10.1186/1471–2105–13–71

10. Yan, Q. Translational Bioinformatics Support for Personalized and Systems Medicine: Tasks and Challenges. *Transl. Med.* **2013**, *03*(02). http://dx.doi.org/10.4172/2161– 1025. 1000e120

CHAPTER 13

THE SHIFTING PARADIGM OF THE EMERGING TECHNOLOGY OF INFORMATION AND COMMUNICATION TECHNOLOGY: CHEMINFORMATICS, TO THE BENEFIT OF SCIENCE

HERU SUSANTO[1,2]

[1]The Indonesian Institute of Sciences, Jakarta, Indonesia

[2]Tunghai University, Taichung, Taiwan

Corresponding author. E-mail: heru.susanto@lipi.go.id

ABSTRACT

The improved emerging technology of information and communication technology (ICT) has dramatically changed data and information processing. The integrated components that collect, manipulate, store, and disseminate data and information provide a feedback mechanism to meet specific objectives.

Technology has significantly contributed to improving our lives and could also participate in creating a new whole dimension of science in order to improve the development and assimilation of human life and capabilities of technology. ICT helps scientists to make new discoveries in science to find solutions to the problems and to gather raw data and molecular design from previous research to synthesize new drugs. A balanced combination of emerging technology with instrumentation will help in advancing the process of synthesis without complication. Cheminformatics, the integration of chemistry and information systems, has proven to be advantageous as it assists chemists to comprehend and gather the features of molecules with certain pharmacological properties through innovation and through the accessibility of databases to measure and analyze the discovery of potential new drugs. This study highlights the emerging technology of ICT as a driver for shifting it to the benefit of science.

13.1 INTRODUCTION

Information systems (IS) are similar to information communication technology (ICT). IS consists of raw data collection (input) that can be processed into an additional value for an organization (output). After being analyzed and processed, it would turn into information into data or facts that could be used to answer questions, solve problems, or to conduct a project. Moreover, IS, also known as software, which is being used in an organization to help organize and analyze databases into useful information, can be used in decision making process.

Information technology (IT) has been around for quite a long time. Essentially, for as long as individuals have been around, IT has been around in that there were always ways of communicating through methods accessible at that point in time. There are four main ages that divide the historical background information of IT. Only the latest age (electronic) and a percentage of the electromechanical age absolutely influence us today. IT comprises three fundamental parts: computational information processing, decision support, and business programming. IT is broadly utilized as a part of the business and the field of computing. Individuals use the terms generically when alluding to different sorts of personal computer (PC)-related work.[13,10,3]

13.2 LITERATURE REVIEW

The use of technology is very crucial in this modern world. The newly improved advanced technology has changed our lives dramatically and made it much easier to anticipate the demand for our needs and wants. IS is practically being used everywhere and it has been evolving continuously in the form of devices such as smartphones. Now, it has been moving to home appliances such as television, which is currently called a "smart television."

IS is an integrated component that collects, manipulates, stores, and disseminates data and information and provides a feedback mechanism to meet objectives for individuals, groups, and organizations. It is a vital processor that every computer needs because without it, IT and ICT will not be able to function properly or follow the instructions given by us. This is because IS is built with software that comes from the manufacturer itself where it has its own function and follows the task given by us.

In today's world, technology has significantly contributed in improving our lives. Now, IS could participate in helping science to improvise the medical methods into a new whole dimension of science in order to improve the development and assimilation of human life and the capabilities of technology. Not only this, but it could also help scientists to conduct practical research to make new discoveries in science to find solutions to problems of that humans have been facing, as such, new remedies are to fight diseases. At present, scientists all around the world are working hard to uncover solutions to what we, the humans, are actually facing, not only in terms of medicine but also in terms of new methods in the field of surgery and technology.

The way in which IS could help science would depend on the characteristics and how it is being used. The development in advanced technology and human capabilities, by using IS software, could help to gather all the raw data and molecular design from the past research to synthesize a new drug. Hence, it would help to discover new medicines or drugs to cure currently incurable diseases. Combining emerging technology with high instrumentation will help in advancing the process without complication. For instance, the research that was being conducted for the past couple of decades was immensely complicated because the researchers had to go through every practical research and book, whereas now we have advanced machinery, technology, information, and the internet to help us.

13.3 DISCUSSION

13.3.1 INFORMATION SYSTEM (IS)

The range of definition is given for this concept; however, IS does not have a specific and uniform definition. N. Winer defined IS as who determines the content of the information gleaned from the outside world in the process of our adjustment to it and adapts it to our senses.[12]

IS is the combination of people's innovation and computers that process or unravel information. ICT that an organization uses as well as the medium by which people interact lead to support for business processes. There is no specific and clear distinction among ISs, computer systems, and business processes. Nowadays, ISs are mainly used to gain maximum benefit by processing data from input to generate IS, a combination of hardware, coordination, and decision-making in an organization. It is also known as a decision-making support system. It is a collective combination of people, procedure, software, database, and device that supports problem-specific decision making.

The main usage of IS in an organization is communication. It allows people to communicate easily; for instance, one of the methods of communication, that is, electronic mail or e-mail, is delivered extremely fast and can be sent and received from devices all around the world that have an internet connection. In addition, the availability of cloud function allows people to make changes, add information, and share with one another in a community cloud. Thus, it makes communication better and faster as most information is now easily accessible, where people could gain the right information at the right time. In order to meet one's needs and wants, it is important to obtain accurate and complete information.

A business process may be difficult as entrepreneurial culture and the degree to which the existing ISs tend to represent the compatibility and application functionality significantly affect a firm's propensity to adopt cloud computing technologies. The discovery supports our abstract development and suggests complementarities between innovation diffusion theory and the information processing view. Industry professionals aid in making more informed adoption decisions in regard to cloud computing technologies in order to support the supply chain.[11,22,2]

Owing to the fact that IS plays a major role mainly in business by creating new products and services, it is possible for managers to use real-time data when making a decision; therefore, information must be relevant and reliable in order to help people in organizations to achieve their goals and perform tasks more effectively, which can then lead to competitive advantage and can make workers' jobs as easier. Messner stated that information is the data of monetary phenomenon and process used in decision-making processes,[1] namely for human resource management, marketing, and administration.

Moreover, IS could store documents, histories, communication records, as well as operational data that could be used in the future for better reference as it could act as useful historical information. It improves the efficiency of a certain organization's operation in order to achieve aims, objectives, goals, and higher profitability. IS needs to be flexible so that it can accommodate a certain amount of variation regarding the requirement in terms of supporting the business process. The impact of IS being flexible enables cost efficiency for the business.

IT tends to be applied within business operations that can save a great deal of time during the fulfillment of daily tasks. Business tends to include knowledge management, artificial intelligence, expert system, multimedia, and virtual reality system. Paperwork is processed immediately, and financial transactions are automatically calculated by IS. Although businesses may view this expediency as a boon, there are untoward effects to such levels of automation. As technology amends, tasks that were formerly performed by human employees are now carried out by computer and ISs. This leads to the elimination of jobs and, in some cases, alienation of clients, thus lack of face-to-face interaction. Unemployed specialists and once-loyal employees may have difficulty in securing future employment.

Therefore, as much as IS could help make tasks easier for an organization, it has certain issues and obstacles that an organization has face. First are the cultural challenges: each country and regional area has its own intellectual awareness or culture as well as its own customs that could significantly affect individuals and organizations involved in global trade, which means that it is a contagious issue whose influence can affect our culture. An organization should also be aware of the best way to approach global demographics, which have profound significance on the global landscape as well as on the profession of globalization.

Second, limitation of language usage, where language differences create an issue, can make it difficult to translate the actual meaning of a conversation for instance. It is due to the fact that meanings from another language may be different, which leads to misunderstandings, irritations, feelings of exclusion, and a sense of inferiority. It is a daily challenge for the speakers who have not spoken English and are trying to communicate in the language of global business. English is mainly used across the world for global conversation. Moreover, on the internet, there are no facial expressions, body language, or other nonverbal cues, which makes communication even more complex.

Third is time and distance challenges where these issues can be difficult to overcome for individuals and organizations involved with global trade

276 Chemical Technology and Informatics in Chemistry with Applications

in remote locations. Moreover, large time differences make it difficult to communicate directly with people on the other side of the world. With long distance, it can take days to get the products or parts from one location to another. Although IS and technology may be able to make communication faster and easier to use for individuals and organizations, it may lead to difficulty in terms of different time zones and distance, where, for instance, a meeting would be difficult to schedule to get a group of individuals from all over the country through video conversation.

Fourth is technology transfer issues, where most governments do not allow certain military-related equipment and systems to be sold in some countries. As access to capital is limited, the capital costs of electronic systems tests are generally higher than those of standard technologies. Moreover, the risks of identifying the existence of new technologies and financing costs will tend to be higher. Moreover, the availability of foreign direct investment is restricted and unevenly distributed around the world. Although many countries are reviewing their trade policies in order to loosen restrictions in terms of the markets, substantial tariff barriers remain an obstacle in the case of imports of external technologies, including energy supply equipment. This limits exposure to energy in terms of productivity, resulting in pressure to improve from foreign competition on national suppliers and hinders early introduction of sustainable energy that can be maintained at a certain level by alternation from abroad where the foreign exchange of restrictions and public revenue deliberation make across-the-board tariff removal difficult.

Lastly is regarding trade agreements—international agreements on the condition of trade in goods and services. Countries often enter into trade agreements with each other. Although it creates a dynamic business climate where business is protected by the agreement, trade agreements may lower government spending instead of government putting funds for better use, such as by increasing the number of experts who could develop local resources who then help local entrepreneurs, instead of the consequence resulting in an increase in job outsourcing, a reduction in tax revenue as without tariff and fees, and a need to have to find a way to replace the revenue.

Moreover, one main problem is that ISs may not function properly, affecting the business processing where the system may conflict with business strategies; the system analysis or design may perform incorrectly and software development may inherit properties such as complexity, conformity, changeability, and invisibility. It can result in system break-down, which can interrupt smooth operations and lead to consumer dissatisfaction. As has been noted, defective ISs can deliver wrong information to other systems, which could create problems for the business and its customers. In other words, ISs are also vulnerable to hackers and frauds.[4],[23]

13.3.2 INFORMATION AND TECHNOLOGY

IT is considered as a subset of data frameworks. It manages innovation part of any data framework, that is, equipment, servers, working frameworks, and programming. A PC-based device that individuals use to work with data and backing the information and data processing is need of an association. It concentrates on innovation and helps in sharing data. A business can utilize IT to create organization database applications which can permit employees to access information at any given moment. They can also use IT instruments to set up networks that permit departments to share information without any problem or wastage of time. With IT, most associations have made a decentralized enlisting structure which unites the entire scope of the business' information in a methodical manner with the objective that it can be accessed and used by any person who needs it. This structure of information is generally a database, which is planned to clearly support the idea of shared information.[24]

The advantages of IT include that it has united the world and has permitted the world's economy to end up a solitary reliant framework. With IT, we can share data rapidly and proficiently, and it can cut down hindrances of phonetic and geographic limits. The world has formed into a worldwide town because of the assistance of data innovation permitting nations such as Chile and Japan that are isolated by separation as well as dialect to impart thoughts and data to one another. Moreover, with the assistance of data innovation, correspondence has additionally become less expensive, faster, and more proficient. We can now correspond with anybody around the world basically by a message informing them or sending them an e-mail for a practically prompt reaction. The web has additionally opened up face-to-face direct correspondence from various parts of the world because of the aides of video conferencing.

In addition, data innovation has electronized the business handling, subsequently streamlining organizations to make it amazingly practical cash-making machines.[8] This, thus, expands profitability which at last offers ascend to benefits that imply better pay and less strenuous working conditions. Data innovation has spanned the social crevice by peopling from various societies to correspond with each other and takes into account the trading of perspectives and thoughts, along these lines expanding mindfulness and decreasing partiality. IT has also made it feasible for organizations to be open 24×7 everywhere throughout the globe. This implies a business can be open at whatever time in any place, making buys from various nations less demanding and more helpful. It additionally implies that merchandises conveyed right to the doorstep with moving a solitary muscle.

In terms of education, IT makes it conceivable to have an online education. Unlike in the past when education was tied to particular limits, now the education sector has changed. With the establishment of online education services, students can learn from anywhere utilizing the internet. This has helped in spreading of vital education materials to all individual, mainly students across the globe. Online education is also being improved by the making of portable application which empowers students' access to educational material through their mobile phones.

In terms of agriculture, IT plays a major part in advancing the agricultural sector. These days, farmers can sell their products right from the homestead utilizing the internet. All they need to do is to create a site to advertise their products; orders will be placed directly by means of the site and the farmers will deliver fresh goods to the consumers once orders have been made. This gets rid of the middlemen who tend to increase the cost of farming products with the aim of making profits. In this case, IT benefits both the farmer and the consumer. The consumer gets the product at a low price while it is still fresh, and the farmer gains additional income.

Somehow, most likely the best point of interest of data innovation is the making of new and intriguing occupations. PC software engineers, systems analyzers, hardware and software designers, and web fashioners are only a portion of the numerous new business opportunities made with the assistance of IT. In addition to that, job posting sites usually use IT as a classification in their databases. The class incorporates an extensive variety of employment across architecture, engineering, and management functions. Individuals with occupations in these areas commonly have a progress on education, mainly in software engineering and/or ISs which they may also possess related industry certifications. Short courses in IT fundamentals can also be discovered online and are particularly helpful for the individuals who want to get some introduction to the field before focusing on it as a career. A career in IT can include working in or leading IT departments, product advancement teams, or research groups.

Headways in data innovation have had numerous huge points of interest on society, and might be the crown gem of our time and indicate the progression of humankind; however, this has not come without its disadvantages to IT or ISs, which may leave individuals to think about whether the great exceeds the terrible. A number of weaknesses of data innovation have been incorporated; while data innovation might have streamlined the business process, it has additionally made occupation redundancies, cutting back and outsourcing. This implies a ton of lower and center-level occupations have been done away with, bringing on more individuals to end up unemployed. Despite the fact that unemployment and occupation are due to IT, it is no place close in the examination as IT has unquestionably delivered the inconceivable amount of employment. Change is tragically constant, and in business terms, you need to move with movement or be abandoned.

Despite the fact that data innovation might have made correspondence speedier, less demanding, and more advantageous, it has additionally purchased along security issues. From wireless sign block attempts to e-mail hacking, individuals are currently agonized over their once-private data getting to be open information. A significant measure of individuals is uninformed of the endeavors substantial organizations go to gather information on individuals and the utilization and offering of this information. By and large, thoughts, for example, online treats, which promote the web client's hobbies, can be seen as something to be thankful for yet one could think about whether faculty data processing is something worth being thankful for in the hands of extensive organizations whose essential premium is to inspire you to spend your well-deserved cash.

On the other hand, industry specialists believe that the web has made professional stability a major issue since alternation continues every day. This suggests that one must be in a steady learning mode, on the off chance that he or she wishes for his or her business to be secured. This disservices of IT or IS has been around since the presentation of all innovation and one must not overlook that life is a ceaseless learning cycle and that you should stick to it or be abandoned. Moreover, as computing systems and capabilities keep growing worldwide, information overload has turned into an undeniably critical issue for many IT experts. Efficiently processing immense measures of information to produce beneficial business intelligence necessitates a lot of processing power, sophisticated software, and human analytic expertise.

As data innovation might have made the world a worldwide town, it has likewise added to one society overwhelming another weaker one. For instance, it is currently contended that the United States impacts how most youthful young people everywhere throughout the world now act, dress, and carry themselves. Dialects too have gotten to be dominated, with English turning into the essential method of correspondence for business and everything else. Loss of dialect and society is never something worth being thankful for, yet IT or IS additionally is impressive at holding learning of society, dialect, and one-of-a-kind practices, so that individuals hold their character or social personality.

Overdependence on innovation where computers and the internet have turned into a fundamental part of this current life, a few individuals, particularly, youths who grow up with it, would not have the capacity to work without it. The internet is conceivably making individuals sluggish, especially with regards to task or venture research as opposed to perusing books in a library, individuals can simply do a Google search. Moreover, the usage of technology in an organization, company, or business decreases the number of hours that a human works at that company. This may even result in some people losing their work because technology is doing it for them. However, this is advantageous for the organizations as their profit will increase because they do not need to pay their workers as much because they are not needed as much.

With the ever-growing variety of social networking sites such as Facebook and Twitter, it is not impossible that the traditional communication skills will be lost; especially, children are always engrossed in these websites because the exchange of information and responsive skills are not important with computers. E-mails and instant messaging have replaced the old tradition of handwritten letters. Although this is advantageous considering time constraints, a personal touch and sense of feeling are lost in comparison to consuming the time to sit down and handwrite a letter.

In terms of health, studies have proved that technology can create an a number of problems with a person's health.[17] Many scientists, doctors, and researchers are worried about potential links between technology and heart problems, eye strain, obesity, muscle problems, and deafness. Waste released from technology can contaminate the environment which not only makes people ill but also harms the environment.[15,16]

13.3.3   IS AND IT SUPPORT SCIENCE

The availability of open access and online chemical databases has made it easier for the people to know more about what is going on in the chemistry world and to be updated with the recent findings. Integrating databases with other resources, including journal writing, was important for advance scientific progress. Enhancement of data and information integration specifically in scientific software system has become an issue of awareness among the chemists and the cheminformatics community for the past few years but the issue has been solved by the development of the Semantic Web techniques.

In the field of sciences, in order to develop a new research, it is always based on the previous findings. Therefore, it is crucial to keep a record of the previous concepts of sciences so that it can be used in the future whether for improvements on the finding or for the benefit of references.[6,14]

The Semantic Web technique itself is the inclusion of machine-processable data in web documents and it is aimed to transform information which has not been structured or only semi-structured into a fully organized web document which will be made accessible both to humans and machines. There are actually three major sections in the Semantic Web technique; they are the Dublin Core, Open Archives Initiative Object Reuse and Exchange (OAI-ORE), and finally Simple Knowledge Organization System (SKOS). In this modern world, IS plays a very important role especially in collecting, organizing, and storing data or information. It is almost impossible for an organization to not have IS department. IS also consists of a series of advanced technology such as the latest software and hardware. In order to achieve information faster and in an organized manner, software such as Microsoft Excel might be of help in such situation. These sophisticated technologies can only be functioned to its full potential by people who have the accomplishment in handling those technologies. Basically, IS is initially created for supporting operations, management, and decision-making in an organization. It also helps an organization in terms of communication networking, processing, and interpreting data.

Science is the body of knowledge of the physical and natural world which often requires the support of technology. The advancement of science is largely maintained by the frequently updated technology. Especially, in the field of research, technology is probably the most significant necessity in ensuring the success of that certain research. Even tools and equipment used in science cannot be developed without the help of technology. The results of a science research which is made possible by the existence of technology are usually used for the benefits of the society such as drug discovery in the medical world.

For instance, drug discovery nowadays requires proper management system and the enhancement of the accessibility of potentially useful data. This can only be achieved by the existence of an IS and IT. If IS and IT were to be combined with the advancement of science, surely they will produce an efficient and more effective finding in any research. This is due to the requirements needed for a research to be conducted, such as the complicated comprehensions of words in chemistry which can only be interpreted using a certain program such as the Semantic Web techniques.[5,19]

Semantic as we know is the study of meanings of words, phrases, signs, and symbols. In chemistry, alone, there are plenty of words, phrases, signs, and symbols which are not familiar to the people who are not related to the field of chemistry. Fortunately, with this newly founded technology, it is easier for the consumers to search for the meaning of a certain word, phrases, signs, or symbols. A Semantic Web technique is specially created for the purpose of smoothing of any chemistry research.

The Dublin Core specifically focuses on definitions of specifications, vocabularies, and best practice for the assertion of metadata on the web. Dublin Core is an initiative to create a digital "library card catalog" for the web and the elements that offer extended categorized information and improved document indexing.

Similarly, OAI-ORE is used when some resources have a meaningful relationship with other resources or are becoming a part of other resources, such as a figure or a table which belongs to another resource of a scientific publication. Another example is when a resource is being associated with another resource, for example, when a review is made, it is related back to its original text of the scientific publication. The automated software system will then manipulate these sources as a whole instead of separating them. This technique can definitely make a certain research or reference a lot more efficient, effective, and less time-consuming.

Then finally, the SKOS is a project that encouraged publication of controlled vocabularies on the Semantic Web. It is also highly dependent on informal methods, including natural language. Other important aspects of IS and IT in terms of supporting science are validity, accountability, and value proposition, to name a few. Validity is deeply important, for instance, in a laboratory environment, an invalid risk assessment could have negative consequences, where endangerment of human life is inclusive. In the case of accountability, an organization or an individual is accountable or responsible for the validity of the information that they provided.

Value proposition depends on individual perspective and also organizational perspective. From an individual perspective, it is less time-consuming and the data or information provided has been standardized so that it is easier to carry out a research. From an organizational perspective, it is actually risky to provide a source of information as an unauthorized access can easily condemn or leak out valuable information on the website if the website has a weak safety system.

Science, innovation, and advancement speak to a progressively bigger classification of exercises which are profoundly associated, however particular. Science adds to innovation in no less than six ways.

1. New learning which serves as an immediate wellspring of thoughts for new mechanical potential outcomes
2. Wellspring of devices and methods for more proficient building outline and an information base for assessment of attainability of plans
3. Research instrumentation, laboratory strategies, and expository techniques utilized as a part of examination that in the end discover their way into configuration or mechanical practices, frequently through the middle of the road disciplines
4. Routine of exploration as a hotspot for improvement and absorption of new human aptitudes and abilities, in the end, helpful for innovation
5. Formation of an information base that turns out to be progressively critical in the appraisal of innovation as far as its more extensive social and natural effects
6. Information base that empowers more productive methodologies of connected examination, advancement, and refinement of new innovations

On the other hand, the innovation affecting science in any event is measured by giving novel exploratory inquiries and methods to address novel and more troublesome investigative inquiries more effectively.

Particular illustrations of each of these two-way connections are stated. Owing to numerous backhanded and coordinate associations in the middle of science and innovation, the exploration arrangement of potential social advantage is much more extensive and more looking so as to differ than would be recommended just at the immediate associations in the middle of science and innovation.

The institute has previously distributed contextual investigations on one of the advancements that has occurred regarding the interest-driven material science research. The studies show that the long timescales in which the procedure between the first disclosures and the advancement of items that use the examination can happen. Four key advances highlighted in these distributions that empowered a number of the innovation-based developments found in the administration areas are portrayed beneath.

The first example is fiber optics where the advancement of fiber-optic advances took into consideration broadband web associations and quick overall correspondence of data, empowering online developments in the administration areas, for example, virtual interfaces, online medicinal services checking, and remote systems administration. The innovation has its roots in material science research by John Tyndall in the 1800s and in later research into photonics.

Second, the utilization of lasers took into consideration both quick correspondence through broadband systems, furthermore for quick information stockpiling, and recovery through compact and digital versatile disc or digital video disc advances. The guideline behind the laser was produced by Albert Einstein and it took over 40 years before the principal obvious wavelength laser was built. Analysts at the University of Surrey are as of now attempting to control quantum course lasers which could be utilized for restorative analysis, for instance, glucose observing for diabetics.

Third, liquid crystal display (LCD) innovation empowers cell phones to be lightweight, low-control utilization, and ease screens. The first exploratory examination that supports LCD innovation was led over 100 years prior, with further advancement in the second half of the most recent century. Basic LCD showcases are found in watches and number crunchers with more intricate shows now in cell telephones, PC screens, and televisions.

Lastly is global positioning system (GPS) where the capacity to precisely decide the position of an article or individual has empowered advancements, for example, web-based robbery following of autos and satellite route. GPS is supported by an extensive variety of material science research, from nuclear timekeepers to the hypothesis of *general relativity*, consolidated with space science and innovation.

13.3.4 CHEMINFORMATICS

Cheminformatics, otherwise called chemoinformatics and compound informatics, is the utilization of PC and instructive procedures connected to a scope of issues in the field of science. These in silico strategies are utilized as a part of, for instance, pharmaceutical organizations during the time spent in medication disclosure.

The utilization of cheminformatics instruments is becoming significantly important in the field of translational exploration from medicinal chemistry to neuropharmacology. Specifically, it requires the investigation of substance data on huge datasets of bioactive mixes. These mixes frame vast multi-target complex systems; it is a drug-target interactome system that brings about an extremely difficult information examination issue. Counterfeit neural network calculations may offer some support by anticipating the collaboration of medications and focusing on central nervous system interactome.[22,21]

Cheminformatics can be simple with the right devices as it begins with the database. It keeps all your basic data and delegate information readily available. In any case, all the more imperative is a cheminformatics framework like CDD Vault. CCD is a hosted database solution for secure analysis, management, and sharing chemical biological data, where it let to intuitively organize the chemical structure and biological structure data, and allows collaboration within the organization internal or external partners through a web interface; it incorporates every part of the instruments important for a simple-to-utilize and end-to-end arrangement. It includes registration, synthetic drawing, looking, and representation, and additionally structure–activity relationship (SAR) and different examinations. CDD Vault even serves as the storehouse for more than 2 million open mixes and examines results as well. CDD Vault gives a complete medication revelation informatics framework, effortless information for action, compound similitude, and selectivity.

It performs SAR examination to recognize vital basic components.

It is the center for obtaining and screening endeavors on the ideal one-of-a-kind arrangement of mixes survey drug suitability with Lipinski rules, Opera lead-like measurements, and other custom descriptors.

CDD Vault makes the testing objectives of cheminformatics less troublesome where commonplace web interface makes cheminformatics more available to learners. In addition, incorporated diagrams offer some assistance with visualizing the relationship in the middle of action and properties.

What is more, as a protected, facilitated cloud application, CDD Vault is a practical arrangement that is ideal for scholastic gatherings, charities, and little organizations. Current facilitated cloud design gives cost investment funds, additionally making it conceivable to fabricate more natural interfaces, better than legacy frameworks. At the point when cheminformatics work includes others, CDD Vault makes coordinated effort simply with implicit correspondence and sharing capacities. The precise cut of the information with just the accomplices is determined. For instance, cheminformatics analysis of organic substituents identifies most common substituents, its calculation of substituent properties, as well as mechanical recognition of drug-like bioisosteric groups.[10,20]

Namely, Enterprise Capability is an industrial strength database with all the benefits of the cloud which is easy to use. The fact is that there is zero footprint web interference run on all major browsers, served from the certified cloud for immediate turnkey deployment. As for the cloud technology, it provides private, multi-tenant architecture that is able to provide affordable security and has demonstrated 99.98% history availability. Moreover, it provides management tools that are easy to track in real-time changes, showing full details of new compounds and data. Moreover, it is customizable where it could define protocols, data fields, preferred graphs, and even chemical registration business rules; however, customization services are available for more complex requests.

The entire combinatorial science is likewise taking into account the idea of substituents, acting for this situation under the name-building squares. The present study concentrates on natural substituents from the perspective of cheminformatics and tries to reply questions about the aggregate number of substituents in known natural science space and the ramifications of this number for the span of virtual natural science space. The portrayal of substituents by ascertained properties is examined as well, including a technique for computing substituent drug resemblance with respect to an examination of the dissemination of substituents in a vast database of medications versus an extensive database of nondrugs. Lastly, an illustration of the application of a vast database of medication such as substituents with computed properties is presented a web-based instrument for programmed distinguishing proof of bioisosteric gatherings.

Cheminformatics is most likely to be beneficial for science field today and equally to the world. The integration of chemistry and IS has proven to be advantageous as it assists chemist to comprehend and gather all the features of the molecules with certain pharmacological properties. This innovation and accessibility of database have helped to measure and analyze the discovery of potential new drug. Moreover, because of the unlimited accessibility in the field of cheminformatics, chemists have the ultimate pleasure to use enormous tools and methods that are advanced, safe, and bioavailable. Along with this, a different view of comprehension and predicting the bioactivity has its strength in partial least-square or genetic algorithms.

Unfortunately, every advantage has its own limitation such as the discovery period of a new drug will take quite some time. The tool and approach of cheminformatics might not have the potential and might not be accurate enough to reveal new discovery and might be risky as well. Moreover, it might not utilize its fullest potential and impacts than we realize. An increase in the cost of drug discovery has brought a rise in the efficiency industry only; hence, it encourages people to conduct a research only if it is profitable. Lastly, it requires a huge capital and maintenance.

13.4 CONCLUSION

The advancement of IT achieved a turning point with the improvement of the internet. Through the course of its improvement, specialists started finding different utilizations for the network and utilization of the technology network around the world. An access to the internet today by people, organizations, and institutions alike has created a worldwide business sector for internet service and has spurned an increase in productivity in the technological communication field. IT is continuously developed in order to enhance today's organization system to be more manageable, productive, and systematic. Constant improvements can possibly create new applications of IT that can affect all the areas of the society which includes the economy, households, government, and private sectors. Therefore, it is important to always be aware of the latest update of the technology in order to use the IS and IT to their maximum potential; it will definitely result in improvement, not just in an organization but also among individuals of the society.

13.5 RECOMMENDATION

Although IS is already beneficial to almost all organizations, it is still open to embrace improvements. One of them is to create a Technology Usage Agreement for the staff of that certain organization. For instance, an organization has the right to control its staff from browsing or visiting inappropriate websites that are known to house viruses such as torrent and file-sharing sites. In addition, the organization could also develop limitation for data or music downloading policy and include terms and conditions for data confidentiality. This is to assure that the organization confidentiality is safely guarded by its staff.

The second improvement that can be made is having a backup plan. It is impossible to know what will happen to an organization in the future; hence, in order to be safe, a backup plan is essential. One solution for this issue is to hire or consult one of the many firms that provide an off-site storage where an organization can keep a record of the key documents and important databases so that when an unexpected disaster happens, the organization can still recover by referring to the backup plan.

The third betterment that can also be made is by setting up a schedule periodic maintenance downtime where the employees of the IS and IT department can manage update, scan for viruses, backup all the data, and fix errors. This is for the prevention of data loss and system malfunction.

Then, the fourth method that can be considered is to contact the internet service provider of an organization. An organization can request to increase its bandwidth. Increasing bandwidth has a lot of advantages for the organization such as making multitasking much easier and it could also reduce application hang-ups caused by slow updating of the software of that certain application.

After that, the organization may also create a comprehensive technology plan. A comprehensive technology plan is where an organization can monitor the updates of technologies nowadays and hence use it for the benefit of the organization. The results of the monitoring of the latest technologies can be evaluated closely and then it can be adopted into the organization to replace aging workstations that are no longer efficient and effective in today's modern world.

Finally, an organization must at least make the effort to create a website which is user-friendly. This can be done by considering software as a service. The organization then can provide many options for functions such as word processing, search engine, contact person or e-mail, and most importantly, customer relationship management function.

KEYWORDS

- drug discovery
- cheminformatics
- compound simulation
- algorithm
- communication technology emerging technology
- gene database

REFERENCES

1. Abbott, M. A.; Messner, R. A. Use of Coordinate Mapping as a Method for Image Data Reduction. *In Intelligent Robots and Computer Vision IX: Algorithms and Techniques;* International Society for Optics and Photonics, UK, February 1991, Vol. 1381, pp 272–283.
2. Almunawar, M. N.; Anshari, M.; Susanto, H.; Chen, C. K. How People Choose and Use Their Smartphones. In *Management Strategies and Technology Fluidity in the Asian Business Sector;* IGI Global, Singapore, 2018a; pp 235–252.
3. Almunawar, M. N.; Anshari, M.; Susanto, H. Adopting Open Source Software in Smartphone Manufacturers' Open Innovation Strategy. In *Encyclopedia of Information Science and Technology, Fourth Edition;* IGI Global, Singapore, 2018b; pp 7369–7381.

4. Bajdor, P.; Grabara, I. The Role of Information System Flows in Fulfilling Customers' Individual Orders. *J. Stud. Soc. Sci.* **2014**, *7*(2), pp. 96-106.

5. Borkum, M. I.; Frey, J. G. Usage and Applications of Semantic Web Techniques and Technologies to Support Chemistry Research. *J. Cheminform.* **2014**, *6*(1), 18.

6. Chaves, C. V.; Moro, S. Investigating the Interaction and Mutual Dependence Between Science and Technology. *Res. Pol.* **2007**, *36*(8), 1204–1220.

7. Jamal, S.; Scaria, V. Cheminformatic Models Based on Machine Learning for Pyruvate Kinase Inhibitors of Leishmania Mexicana. *BMC Bioinf.* **2013**, *14*(1), 329.

8. Leu, F. Y.; Ko, C. Y.; Lin, Y. C.; Susanto, H.; Yu, H. C. Fall Detection and Motion Classification by Using Decision Tree on Mobile Phone. In *Smart Sensors Networks;* Elsevier, 2017, pp 205–237.

9. Liu, J. C.; Leu, F. Y.; Lin, G. L.; Susanto, H. An MFCC-Based Text-Independent Speaker Identification System for Access Control. *Concurr. Comp. Pract. Exp.* **2018**, *30*(2), pp. 42-55.

10. Loging, W.; Rodriguez-Esteban, R.; Hill, J.; Freeman, T.; Miglietta, J. Cheminformatic/Bioinformatic Analysis of Large Corporate Databases: Application to Drug Repurposing. *Drug Discovery Today Ther. Strategies* **2012**, *8*(3), 109–116.

11. McGarry, M. P.; Reisslein, M.; Maier, M. Ethernet Passive Optical Network Architectures and Dynamic Bandwidth Allocation Algorithms. *IEEE Commun. Surv. Tutor.* **2008**, *10*(3), pp. 103-115.

12. Power, D. J.; Sharda, R.; Burstein, F. *Decision Support Systems;* John Wiley & Sons, Ltd.: USA, 2015.

13. Susanto, H.; Almunawar, M. N.; Leu, F. Y.; Chen, C. K. Android vs iOS or Others? SMD-OS Security Issues: Generation Y Perception. *Int. J. Technol. Diffus.* **2016**, *7*(2), 1–18.

14. Susanto, H. (2017). Electronic Health System: Sensors Emerging and Intelligent Technology Approach. In *Smart Sensors Networks*; pp. 189–203.

15. Susanto, H.; Chen, C. K. Information and Communication Emerging Technology: Making Sense of Healthcare Innovation. In *Internet of Things and Big Data Technologies for Next Generation Healthcare;* Springer: Cham, 2017, pp. 229–250.

16. Susanto, H. Smart Mobile Device Emerging Technologies: an Enabler to Health Monitoring system. In *High-Performance Materials and Engineered Chemistry;* Apple Academic Press, Canada, 2018a; pp. 241–264.

17. Susanto, H. Cheminformatics—The Promising Future: Managing Change Of Approach Through ICT Emerging Technology. In *Applied Chemistry and Chemical Engineering, Volume 2: Principles, Methodology, and Evaluation Methods;* CRC Press, Canada, 2018b; p. 313.

18. Susanto, H. Biochemistry Apps As Enabler Of Compound And DNA Computational: Next-Generation Computing Technology. In *Applied Chemistry and Chemical Engineering, Volume 4: Experimental Techniques and Methodical Developments;* CRC Press, Canada, 2018c; p. 181.

19. Susanto, H.; Chen, C. K. Macromolecules Visualization Through Bioinformatics: An Emerging Tool of Informatics. In *Applied Physical Chemistry with Multidisciplinary Approaches;* CRC Press, Canada, 2018a, 21.

20. Susanto, H.; Chen, C. K. Informatics Approach and Its Impact for Bioscience: Making Sense of Innovation. In *Applied Physical Chemistry with Multidisciplinary Approaches;* CRC Press, Canada, 2018b.

21. Susanto, H.; Chen, C. K.; Almunawar, M. N. Revealing Big Data Emerging Technology as Enabler of LMS Technologies Transferability. In *Internet of Things and Big Data Analytics Toward Next-Generation Intelligence;* Cham: Springer, 2018; pp 123–145.

22. Tran, S. T.; Le Ngoc Thanh, N. Q. B.; Phuong, D. B. Introduction to Information Technology. In *Proceedings of the 9th International CDIO Conference (CDIO)*, 2013.

23. Wu, Y.; Cegielski, C. G.; Hazen, B. T.; Hall, D. J. Cloud Computing in Support of Supply Chain Information System Infrastructure: Understanding when to Go to the Cloud. *J. Supply Chain Manage.* **2013**, *49*(3), 25–41.

24. Zhang, X.; Zheng, N.; Rosania, G. R. Simulation-Based Cheminformatic Analysis of Organelle-Targeted Molecules: Lysosomotropic Monobasic Amines. *J. Comput. Aided Mol. Des.* **2008**, *22*(9), 629–645.

CHAPTER 14

HIGH-PERFORMANCE GRID COMPUTING FOR CHEMINFORMATICS

HERU SUSANTO[1,2,*], FANG-YIE LEU[2], and CHING KANG CHEN[3]

[1]*The Indonesian Institute of Sciences, Jakarta, Indonesia*

[2]*Tunghai University, Tunghai, Taiwan*

[3]*School of Business and Economics, Bandar Seri Begawan, Brunei*

Corresponding author. E-mail: heru.susanto@lipi.go.id

14.1 BACKGROUND

The term "grid" has emerged in the mid-1900s to denote a proposed distributed computing infrastructure which focuses on large-scale resource sharing, innovative applications, and high-performance orientation (Foster et al., 2001). Grid concept is inspired by a genuine and particular issue—the coordinated resource sharing and problem-solving of energetic, multi-institutional virtual organizations (VOs).

The sharing is not mainly a file exchange but relatively a direct access to computing resources, software, storage devices, and other resources with a compulsory, highly controlled sharing rule which defines clearly and cautiously just what is shared, who is allowed to share, and the circumstances under which sharing occurs. A set of individuals and/or foundations defined by such sharing rules forms what we call a VO.

The earliest definition of a grid emerged in 1969 by Len Kleinrock was as follows:

"We will probably see the spread of 'computer utilities', which, like present electric and telephone utilities, will service individual homes and offices across the country."

14.2 INTRODUCTION

The widespread phenomenon of resource sharing across geographical barriers is made possible with advanced computer technologies—a form of this would be grid computing.

14.2.1 WHAT IS GRID COMPUTING?

Grid computing, in general, refers to a computational system of resource sharing between multiple users of different communities on a local, national, and global scale with the emphasis on the heterogeneity of platforms, hardware and software architectures, and computer languages with common protocols and middleware for uniformity.

For a system to be defined as a grid, there would be a need to fulfill the following distinct guidelines, known as the three-point checklist, as introduced by Ian Foster (2002):

1. In coordinating the resources, there exist different users in separate domains of control as opposed to a centralized control unit.
2. Protocols and interfaces by which the system has been built from are general-purpose and open in nature and are standardized.
3. Possesses nontrivial transmission properties which allow resource components to be co-allocated and coordinated to deliver varying qualities of service in relation to response time, availability, and security; which in turn, help meet the different demands from grid users.

14.2.2 CHARACTERISTICS OF GRID COMPUTING

14.2.2.1 LARGE SCALE

A grid form of the system should be flexible enough to withstand numerous numbers of resources without potentially disrupting or decreasing the performance standard of the grid.

14.2.2.2   GEOGRAPHICAL DISTRIBUTION

The various control centers and components of a whole grid should be located at varying locations of differing distances.

14.2.2.3   HETEROGENEITY

A grid would be composed of different forms of resources; software and hardware, such as scientific instruments, software programs and components, computers, supercomputers, and networks.

14.2.2.4   RESOURCE SHARING

Different users and organizations are able to gain access to resources on a grid; disregarding the original ownership of the resources.

14.2.2.5   MULTIPLE ADMINISTRATIONS

Restrictions can be imposed by individual organizations on the degree of accessibility of their resources belonging to them by forming security as well as administrative policies.

14.2.2.6   TRANSPARENT ACCESS

Different components of a grid would integrate and cooperate to form a single computer.

14.2.2.7   DEPENDABLE ACCESS

Grids are to meet the standards of quality of service (QoS) in terms of service delivery, to facilitate dependable, predictable, sustained, and high-performance levels.

14.2.2.8 CONSISTENT ACCESS

Heterogeneous resources which form the grid would be streamlined with inbuilt features of standardized protocols and interfaces.

14.2.2.9 PERVASIVE ACCESS

The grid should be adaptable to a dynamic environment where instances of resource failure are common in terms of providing access to resources to respective users, as well as in achieving maximum performance from resources obtained.

14.2.3 CLASSIFICATION OF GRID SYSTEMS

Initially, grid computing systems were classified into just two categories—computational grid and data grid. However, other grids were added into the classification, which are as follows.

14.2.3.1 COMPUTATIONAL GRID

Computational grid is defined as systems that control machines of an administrative field in a "cycle-stealing" method to have higher computational capacity than the capacity of any constituent machine in the system. The focus of this grid lies on the optimizing execution time of applications that requires a great number of computing processing cycles.

14.2.3.2 DATA GRID

Data grid indicates the systems that provide a hardware and software infrastructure for synthesizing new information from data repositories that is circulated in a wide-area network. This grid provides the answers for large-scale data management issues.

14.2.3.3 SERVICE GRID

Service grid is referred to as systems that provide services that are not provided by any single local machine. This category is further divided as on-demand, collaborative, and multimedia.

14.2.4 GRID APPLICATIONS

The different types of computing can be categorized according to the main challenges that they present from the grid architecture point of view. The types of computing can be concluded as follows.

14.2.4.1 DISTRIBUTED SUPERCOMPUTING

Distributed supercomputing allows applications to use grids to aggregate computational resources in order to reduce the completion time of a job or to tackle problems that cannot be solved on a single system.

The challenges are as follows:

- The need to co-schedule scarce and expensive resources
- The scalability of protocols and algorithms to tens or hundreds of thousands of nodes
- The design for latency-tolerant algorithms
- Achieving and maintaining high-performance computing (HPC) across heterogeneous systems

14.2.4.2 HIGH-THROUGHPUT COMPUTING

High-throughput computing is used to schedule large numbers of loosely coupled or independent tasks with the goal of putting unused processor cycles often from idle workstations to work. Chip design and parameter studies are general applications of this type of computing.

14.2.4.3 ON-DEMAND COMPUTING

On-demand computing uses grid capabilities to couple remote resources into local applications in order to fulfill short-term requirements. These resources cannot be cost-effective or conveniently located and it may be computation, software, data repositories, specialized sensors, and so on. The challenging issues derive primarily from the dynamic nature of resource requirements and the potentially large populations of users and resources and these include resource location, scheduling, code management, configuration, fault tolerance, security, and payment mechanisms.

14.2.4.4   DATA-INTENSIVE COMPUTING

Data-intensive computing analyzes and treats information and data which are maintained in geographically distributed repositories, digital libraries, and databases, and aggregated by grid capabilities.

The challenge is the scheduling and configuration of complex, high-volume data flows through multiple levels of hierarchy.

14.2.4.5   COLLABORATIVE COMPUTING

Collaborative computing is concerned primarily with enabling and enhancing human-to-human interactions. Challenging issues from a grid architecture perspective are the real-time requirements imposed by human perceptual capabilities and the rich variety of interactions that can take place.

14.2.5   BENEFITS OF GRID COMPUTING

There are numerous benefits of the grid computing system. More importantly, grid computing performs a huge role in the business industry as this is a business model which is evidently beneficial for the businesses. Most businesses nowadays require collaboration and application of resources and policies. Hence, most businesses consider grid computing as a necessity in order to run their business efficiently and effectively. Although there are quite a number of benefits of this model, there are four main distinct benefits of the grids, which are their performance and scalability, resource utilization, management and reliability, and visualization. These benefits will be further explained in the later sections.

14.2.5.1 PERFORMANCE AND SCALABILITY

The performance and scalability of the grids can affect the whole condition and upbringing of a business. For instance, the normal time taken for a business to launch its product to the market is usually time-consuming and lengthy. This is where the performance and scalability of the computational grid take place, as it offers the luxury of sharing infrastructure which can increase and enhance the capabilities through the scalability of the network. As a result, this can save the time taken for the whole process and this will further lead to large cost savings of the business. This, in general, improves the well-being of the industry as a whole.

14.2.5.2 RESOURCE UTILIZATION

The other advantage of grid computing is the ability of fully utilizing the resources that are shared. Resource sharing is one of the examples of resource utilization, where the resources owned by different organization are flowing through the grid and allows the other organizations to access and utilize the information obtained. Enterprises are further appealed by grid computing due to the benefits of sharing the resources not only locally but also geographically. In other words, the resources can be shared internationally regardless of the location of the organization(s). This promotes efficiency to the enterprises and most importantly, reduces the cost of the process.

14.2.5.3 MANAGEMENT AND RELIABILITY

As the ownership of the computational grid can be managed by more than one administration, the grids allow multiple managers for the resources shared. Each of the organization has to manage the resources shared with the appropriate and suitable security as well as the organizational policies so that the other users and organization can access and use them. The four types of access for the resources are transparent access, dependable access, consistent access, and pervasive access. Transparent access refers to the grid being a single virtual computer. On the other hand, dependable access ensures the establishment of the conveyance of the services strictly fulfill the criterion of QoS. This certifies the reliability of the delivered service and the sustained quality of the high performance. Consistent access conceals the heterogeneity of resources while still permitting its versatility, which attracts more enterprises into grid computing. Pervasive access is where the resources granted to be shared have to cater to the dynamic environment of other users and organizations. This ensures the performance of the grid computing to be at the highest level to reduce resource deniability.

14.2.5.4 VIRTUALIZATION

As enterprises are broadening, there is a rapid progression in the usage of mergers and acquisitions due to which heterogeneity is becoming more and more inescapable. Heterogeneity involves both the software and hardware of the resource host to diversify from the range of data, files, software components, computers, supercomputers, network, and so forth, depending on the types of grids that the enterprises use. The network gives virtualization of heterogeneous resources bringing about the better administration of these assets. However, visualization does not resolve the issues and concerns of several factors such as application engineering, security, licensing, and manageability; nonetheless, these concerns can be reduced by research consideration and action. One of the examples of these concerns is licensing. As stated by Chakrabarti,[2] "alternative licensing schemes are required for grid based applications." in order to solve or reduce the concerns raised by virtualization.

14.2.6 STAGES OF GRID EVOLUTION

Grid computing goes through three different generations or stages of evolution and these are concluded as follows.

14.2.6.1 THE FIRST GENERATION

The first generation of grid computing was marked by an early meta-computing environment which emerged in the mid-1990s. In general, these early meta-computing projects have the objective of providing computational resources to a range of high-performance applications.

There are two representatives from this generation which are Factoring via Network-Enabled Recursion (FAFNER) (de Roure et al., 2003) and Information Wide Area Year (I-WAY) (Foster et al., 1997).

14.2.6.1.1 Factoring via Network-Enabled Recursion

FAFNER is referred to as a universal framework that can work at any stage with a web server. In general, the applications of FAFNER can be divided into independent segments, which are accomplished in parallel with each computing resource without the need of a fast interconnect. Its clients are low-end computers and it relies on a group of human involvement to distribute and gather computing results.

14.2.6.1.2 Information Wide Area Year

The I-WAY is intended to manage a variety of different high-performance applications that normally require a fast interconnect and powerful resources at numerous supercomputing centers. The experiences and software developed as part of the I-WAY project have been fed into the Globus venture.

14.2.6.2 THE SECOND GENERATION

In the second stage of grid evolution, there was a growth of core software for the grid of which from providing committed services for large computationally intensive HPC to the more generic and open deployment of middleware.

Middleware is considered to be the layer of software between the operating system and applications, providing a diversity of services needed by an application to function properly. In a grid, the middleware is utilized to hide the heterogeneous nature and present users and applications with a homogeneous and faultless environment by providing a set of standardized interfaces to an assortment of services.

However, there are three main problems that must be tackled:

1. Heterogeneity: A grid involves a range of resources that are varied in nature and might extend frequent administrative domains across a potentially worldwide expanse.

2. Scalability: A grid might develop from a small number of resources to millions. This elevates the issue of possible performance degradation as the size of a grid increases. As a result, applications that need a large number of geographically located resources must be intended to be latency-tolerant and exploit the locality of accessed resources.

3. Adaptability: In a grid, a resource failure is the regulation, not the exception. In reality, with so many resources in a grid, the possibility of some resource failing is naturally high. Resource managers or applications must adapt their performance vigorously so that they can remove the maximum performance from the existing resources and services.

Based on this software, a range of accompanying equipment and utilities is created, providing higher-level services to users and applications, and spreading resource schedulers and brokers as well as domain-specific user interfaces and portals.

14.2.6.3 THE THIRD GENERATION

The third generation is also known as the service-oriented computing. In this phase of grid evolution, there has been a union between grid computing and web service technologies. Therefore, it moved to a more service-oriented approach that reveals the grid protocols using web service standards.

For instance, the open grid service architecture structure is the convergence of web services and grid computing and it carries the establishment, preservation, and application of groups of services continued by VOs. The services have more qualities, simply interactive, and are meta-enabled. Therefore, these services are more adopted in the e-science infrastructure.

In conclusion, the stages of evolution of the grid have been a continuous process and there may perhaps be another stage that will come into sight after the service-oriented computing.

14.3 DISCUSSION

In the following paragraphs, certain issues will be discussed to clarify more about grid computing as a topic of interest. First, the correlation between cloud computing and grid computing will be discussed—to separate the misconception of the facts as these two technology forms are often thought to be closely similar to one another. Followed by the development of a new form of grid, the future for grid computing as well as the significance of grid systems in relation to high-performance computing to society are discussed.

14.3.1 GRID COMPUTING VERSUS CLOUD COMPUTING

Grid computing and cloud computing are known to be correlated with one another, which leads the public to be easily mistaken that these two business models have a similar functioning process. However, the common belief is presented to be incorrect, despite the connections between cloud computing with the recently established grid computing paradigm, as well as other networks such utility computing, cluster computing, and distributed computing systems. Both of these computing systems are networks that abstract processing tasks and provides a more user-friendly interface as they simplify the lengthy and complicated process while using the internet for access. The general differences between grid computing and cloud computing are mostly categorized through their main characteristics and also their functions. Cloud computing is a business model that focuses on several specific functions, such as providing services through the servers that paying users can access and share through the internet; whereas, grid computing focuses on joint and distributed shared resources with the objective of enhancing computational power in the fields. Figure 14.1 shows the diagrams for both business models.

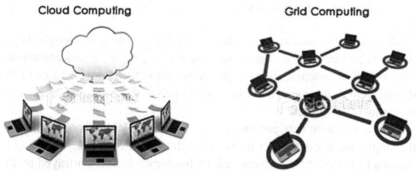

FIGURE 14.1   Grid computing versus cloud computing.

14.3.1.1   CLOUD COMPUTING

Cloud computing services are classified out into three different sections: Infrastructure-as-a-Service (IaaS), Platform-as-a-Service (PaaS), and Software-as-a-Service (SaaS). Conversely, cloud computing is further separated into five layers which are evidently more comprehensible than the previous three sections stated above. These five layers are clients, applications, platform, infrastructure, and servers.

The three sections of cloud computing services serve various purposes and functions. IaaS is responsible for distribution of massive computer resources and capacity of processing, network, and storage. As for PaaS, it serves as a bridge for hardware and the application as it mainly abstracts the infrastructures and is a supporting tool for utilization system interface to cloud applications, while, SaaS performs as a replacement for the software on your computer, such as applications and so forth.

Deployment models characterize the purpose of the cloud and also the natural environment of the cloud. The four deployment models are:

- Public cloud: Open to and accessible by the public, commonly used by large institutes, and owned by organizations vending cloud services.
- Private cloud: Limited access as this infrastructure is solely for specific organizations. This infrastructure can be operated by the organization or a third party.
- Hybrid cloud: A mixture of both public and private.
- Community cloud: An infrastructure which serves for a particular function, such as the organization's mission, strategies, security, administrative consistency needs, and so forth. Managed by a third party or organization(s).

A commonly known example of cloud computing is Google Docs, which is an application that allows users to create, edit, and collaborate with others on documents. Google Docs fulfills the characteristics of the deployment models stated above as it is open to the public and it can also be exclusively limited.

The connection between the services provided by cloud computing and the deployment model can be seen by the statement made by the U.S. National Institute of Standards and Technology which is portrayed by the diagram in Figure 14.2.

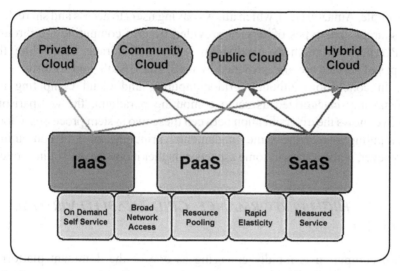

FIGURE 14.2   The U.S. National Institute of Standards and Technology Cloud Computing Definitions.

14.3.1.2   COMPARISON BETWEEN GRID COMPUTING AND CLOUD COMPUTING

This section consists of several lightly detailed comparisons between these two business models. The workflow management of computational grid is dependent on one physical hub, while for cloud computing, an example of its workflow management is Amazon Elastic Compute Cloud (E2). Another key factor that can be compared between these two systems is the level of abstraction. The level of abstraction task process differs in which the level for cloud model is high, on the other hand, the abstraction level of grid model is the opposite to cloud. Additionally, both models are suitable for multitasking; however, cloud computing possesses a higher degree of storage scalability in comparison to the normal level of scalability for the latter. Virtualization is a foundation of cloud computing, whereas this function is not offered in grid computing. Grid model uses a domain name system, which uses the hierarchical decentralized naming system to access the portal, while cloud runs on IP address identification. In the case of ownership for these models, grid computing authorizes multiple ownership as they allow other organizations to access and use their resources; conversely, cloud model is based on single ownership, for example, Amazon (E2), which allows paying users to access and share their resources. The types of service provided by grid computing are through CPU, network, memory, bandwidth, device, and storage. Meanwhile, the types of service are also provided by IaaS, PaaS, SaaS, and so forth.

In conclusion, although grid computing and cloud computing are mistakenly categorized as same computing paradigm, the comparisons above shows the diversification between these two system processes. Cloud computing shares the same fundamental principles of grid computing; however, it also varies in some aspects, whether it downscaled or improved.

14.3.2 HIGH PERFORMANCE: GRID-ENABLED VIRTUAL SYSTEM

Grid computing is rapidly emerging as one of the dominant platforms for wide-area distributed computing.[1] The purpose of grid computing is to provide an environment for sharing and problem-solving in VOs. In the current situation, most of the grid deployments focused on the data-intensive application where the processing was done in a significantly large data sets, and also, costly. The application requires data that is largely distributed in the various storage systems. There is a crucial need to access the remote data with "near-local" performance for scheduling and managing the execution application.

One of the grid's goals is to provide the ability to share and use the data stored in the system as easily as if it was located on the same computer. However, these goals are hard to achieve because it is difficult to use and maintain such environment.

The fundamental problems are:

1. The existence of many administrative domains: The dominance of a particular administrative might lead to restriction or difficulties in accessing the data such as sharing or sending the files.
2. Different storage system: It can lead to having difficulties in finding the data saved, some cannot open the location of the data saved.
3. Different data transfer middleware: The difference in the data transfer middleware may interrupt the data that is being sent.
4. Protocols in grid environment: Due to the protocol in the grid system, difficulties such as changing the grid system need to be

made. For example, if there are changes in the original file and when someone else needs to open it, the person has to use other grid system and he/she need to upgrade the grid system.

Thus, Grid-enAbled Virtual file sYstem (GRAVY)[7] is created, which also facilitates the collaborative sharing of data in the grid. The requirement for the grid virtual system is that it needs a system that serves the random interfaces of the resources and also provides common interfaces. Moreover, the system needs to provide transparency in the data accessibility in the grid environment and also for the user to find the files. For example, the user does not need to know where resources are allocated in order to use them. The system must be able to automatically recover and complete the task desired which is failure transparency.

GRAVY has the following features:

1. Location transparency: It allows users to access the data although it is geographically distributed without the user knowing where the data is originally located.
2. Protocol transparency: It provides a universal data transfer architecture without causing difficulties to the users. As a result, data can be shared easily and in a uniform way.
3. Extensibility: It is not restricted to one protocol and allows other protocols to be added as it evolves to a set of wrappers interfaces.

14.3.2.1 OVERVIEW OF FILE PROTOCOLS

14.3.2.1.1 HyperText Transfer File Protocols

By HyperText Transfer File Protocols, the user makes a Transmission Control Protocol (TCP) connection to the server, send a short-to-long string containing command with few parameters, and then receive the response under the same connection.

14.3.2.1.2 File Transfer Protocols (FTP)

File Transfer Protocol (FTP) is vulnerable because the password is transmitted clearly as data. It also requires making a copy of the data at the remote machine; if the original data has been changed or erupted, new version needs to be updated in the machine.

14.3.2.1.3　Grid FTP

It improves the existing FTP with new features: larger volume, fast data transfer, also support data transfer by TCP streams; thus, it improves the bandwidth over using the single TCP stream.

It also solves the privacy problem with FTP by encrypting both the password and data. Thus, it provides high performance. The data can be accessed in various ways for example by using for example blocked or striped. Part or all the data may be accessed, thus removing the disadvantages of FTP.

Thus, GRAVY solves the problems of protocols in two ways:

1. At the server side: It allows users to use their preferred FTP and interact with GRAVY. Also, it allows GRAVY to be easily and flexibly installed along with the user's need.
2. At the remote side: GRAVY support a large number of existing file system.

14.3.2.2　DATA TRANSFER IN GRID-ENABLED VIRTUAL FILE SYSTEM

Figure 14.3 presents the sequence diagram for a transfer request in asynchronous mode. In this case, the user is responsible for controlling the transfer process and not the TransferManager.

The data transfer in GRAVY consists of following steps:

1. The client uses the asyncCopyTo () operation of GridFile interface to get the FileTransfer file, which is the Runnable interface to be a thread.
2. The client receives the files.
3. GRAVY launches a new thread to execute the file transfer; so, the user is not locked while waiting for completion.
4. In order to check the status of the file transfer, the user can call the get Status () to get the status of the file transferred.

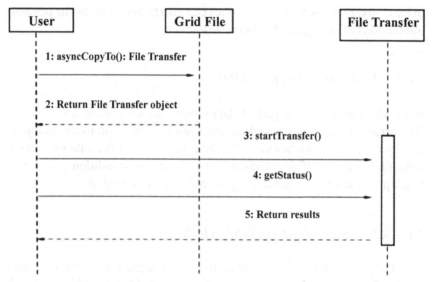

FIGURE 14.3 The sequence diagram for the execution of a transfer request in a synchronous mode.

Source: Adapted from ref [8].

14.3.3 FUTURE OF GRID COMPUTING

The evolution of grid computing can be seen both in data management and execution management in the grid. In the data management, the problem of data management in the grid environment is solved by a service-oriented framework that supports the control of data movement. However, for execution management, the algorithms and strategies are presented and scheduled for a scientific application that had been formulated. Thus, due to these, it raised interesting questions and issue and thus further research is required.

14.3.3.1 SCHEDULING POLICIES FOR DATA MOVEMENT

There are many areas that can be improved in scheduling policies to achieve the best access to data and performance characteristics. In order to achieve the goal of the best performance, the services such as networking service, job scheduler, and so forth, need to work together collaboratively in harmony. Better scheduling policies can also be achieved by the possible interactions of storage and network resources.

14.3.3.2 NETWORK PROFILING

Network is an important part of data movement scheduling decisions. In the current situation, the scheduler does not pay attention to the changing quality of the network at execution time. However, if the data movement scheduler is aware of the changing quality network condition and is able to adapt to the environment, it is possible to achieve high performance.

14.3.3.3 INTEGRATING GRID SERVICES

The important part of the research is to integrate data management service into a global grid service environment. One possible approach is to integrate grid service into the portlets and using a portlet-based portal. A portlet is a web component that generates markup, for example, HTML, XML. Each service of the portlet is connected to one or more grid services.

14.3.3.4 DEPLOYING THE MIDDLEWARE IN A LARGER ENVIRONMENT

Security issue is one of the reasons that restricts the deployment of the middleware across different institutes on the large scale. The research in the future would be able to deploy the middleware in a larger environment with a goal to evaluate the approach in more realistic view.

14.3.3.5 IMPLEMENTING DATA-INTENSIVE SCHEDULING

Usually, in a data-intensive application, a computational grid environment requires large data transfers, which is a costly operation in general. Thus, taking the issue of cost into account is mandatory in order to achieve cost-effective and efficient scheduling of data-intensive application on grid environment.

Several strategies have been proposed in order to reduce the cost of operation of data-intensive applications scheduling. For example, the "bandwidth-centric" strategies view model grid system as a tree structure and consider the optimal solution where tasks are allocated to nodes so that fastest communication time is achieved.

14.3.3.6 INTEGRATING SEMANTIC TECHNOLOGIES

There are a lot of advantages due to the convergence of semantic technology and grid computing. The integration of semantic technology into the web computing raises the level of description, for example, the task-achieving character and capabilities. Thus, due to this, semantic technology needs to be integrated into the web service to provide support, for example, in service recognition, configuration, comparison, and also automated composition.

14.3.3.7 MANAGING GRID WORKFLOW

There is a persistence increase in the usage of grid workflow due to the ability to compose complex applications in a grid environment. The integration of both semantic technology and grid computing provides standards and methods to describe the knowledge of the service. Thus, it also facilitates the automated composition of the service into a larger area and negotiates workflows. In order to support these goals, portal technologies should be researched in detail to discover available grid service, extracting service semantic description, and reduce the composition of the time workflow, especially for the nonexpert users.

14.3.4 IMPACT OF GRID COMPUTING AND HIGH-PERFORMANCE COMPUTING (HPC) TOWARD BUSINESS AND SOCIETY

It has been said that grid computing has been developing and being invested in the past decade by the public research community, the European Commission, and the governments in data supporting infrastructures, according to an article of European Organization for Nuclear Research.

The rise of grid computing has a remarkable impact on both academic and increasingly on businesses due to its ability to create VOs and enterprises for sharing resources for solving a number of problems in terms of sciences and the economy.

Grid computing is much more powerful and flexible than any other ordinary computing. The application of grid computing can improve the accuracy of retrieving data for the organizations so as to receive it at a faster rate, for example, Intergovernmental Panel on Climate Change reports information about the climate change of the Earth from climate scientists. The amount of data is enormous; therefore, grid solution is used to store the information for the global community to analyze and access it.

Another example of an organization that uses grid computing is Bitcoin. Bitcoin allows anyone to make payments to other people globally without the operations of a central authority. Bitcoin network collects transactions into a list known as a block. It is the miner's job to confirm the transactions and record them into a general ledger. Bitcoin was originally designed to be resourceful but it is difficult to maintain the number of blocks found by the miners so the blocks must consist of a proof of work to be valid. Therefore, it provides a smart way to issue the currency and creates an incentive for more people to mine due to its safe security system.

However, the data sets are difficult to produce satisfying requirements due to its process with low probability; as a result, a number of trial and error are required. Therefore, it is time-consuming to generate the data from grid computing. For example, Bitcoin's difficulty to recalculate the number of blocks generated will be in exactly 2 weeks with an average of one block every 10 min.

HPC is a supercomputing that performs at the highest operational rate for computers which use parallel computing for running advanced applications efficiently, which are reliable and inexpensive for the companies, industries, and other organizations. Two examples of an organization that uses HPC are Folding@home and the military of the US Government.

HPC is suitable for the organizations that want their information to be more accurate and secure which consist of one huge data. Folding@home is one of the examples that uses HPC, which was sponsored by Stanford University under a program of Pande Lab. In this, the scientists used HPC to solve the mysteries of the process of protein folding which helps the body to break down food into energy and fight diseases, but it cannot carry out these functions if it is not folded. If it is not folded correctly, it can be known as misfolding which can cause health consequences regarding diseases such as Alzheimer's, Parkinson's diseases, and cancer. With the acknowledgment of protein folding, it could lead to the development of medical treatment.

Researchers at the Los Alamos National Laboratory (LANL) in New Mexico used microprocessors developed for the PlayStation 3 (PS3) to power the fastest supercomputer called the Roadrunner, which was used successfully with the protein folding research. The computational power could make its own virtual calculations both quicker and efficient to handle a number of problems at once. This machine can help to model the physical reactions in the complex situations such as climate change or radioactive fallout simulations. The military of the U.S Government uses the PS3 supercomputer to assimilate virtual situations based on player choices, making a player's screen behave more like the world, as stated by LANL.

However, HPC is only suitable for an organization that does not want its information to be leaked as it is difficult to be maintained and is expensive.

14.4 CONCLUSION

In conclusion, grid computing has a long-running history, dated back from the 1990s. The model of grid computing was merely an idea which was virtualized and developed. The invention of grid computing leads to a shocking discovery of the benefits and ease of the process for the users, more specifically enterprises, as grid computing is a booming beneficial business model. Throughout the years, grid computing faced ups and downs with the creation of almost similar computer paradigm, which is known as cloud computing. The application of cloud computing instantly spread widely and rapidly throughout the years. However, it is discovered that cloud computing and grid computing are indeed not similar, and both business models are useful according to the types of organizations, as grid and cloud offer different services for the users.

Grid computing undergoes three stages of evolution since the 1990s and these stages show that grid computing is improving year by year. Therefore, this evolution of grid computing will never stop because it is a continuous process and there could be the next generation of grid computing after the third one.

Although grid computing has many competitors and it is quite similar to cloud computing, it depends on the user preference. Moreover, grid computing continuously undergoes many research and developments, for example, in reducing the cost with big data transfer to increase the efficiency.

High performance can be a part of the grid computing; as a result, if any company or an organization would want to use HPC, they should save costs since the computers can be dispersed geographically. These are commonly used within the academic community and for economic uses.

Therefore, grid would dominate the computing area because of the persistent and continuous development made for the users.

KEYWORDS

- virtual organizations
- high-performance orientation
- high-performance computing
- factoring via network-enabled recursion
- data-intensive computing

REFERENCES

1. Capello, F.; Djilali, S.; Fedak, G.; Herault, T.; Magniette, F.; Neri, V.; Lodygensky, O. Computing on Large-Scale Distributed Systems: Xtremweb Architecture, Programming Models, Security Tests and Convergence Grid. *Future Gener. Comput. Syst.* **2005**, *21*(3), 417–437.
2. Chakrabarti, A. *Benefits of Grid Computing: Grid Computing Security*; Infosys Technologies Limited: Bangalore, India, 2007.
3. Generations of Grid (n.d). https://gridaccess.wordpress.com/generations-of-grid/
4. International journal of information and computation technology (IJICT) (n.d). http://www.irphouse.com/comp/ijict.htm
5. Karanikolaou, E. M.; Bekakos, M. P.; Milovanovic, I. Z.; Milovanovic, E. I. Grid Computing: Towards Building Supercomputing Platforms. University of Georgia, USA, 2006.
6. Magoulès, F. *Fundamentals of Grid Computing: Theory, Algorithms and Technologies*; CRC: Florida, FL, 2010.

7. Nguyen, T.-M.-H.; Magoules, F.; Revillon, C. GRAVY: Towards Virtual File System for the Grid, In *Proceedings of Advances in Grid and Persuasive Computing*, 2007, pp 567–578.
8. Nguyen, T.-M.-H.; Magoules, F.; Yu, L. *Grid Resource Management: Towards Virtual and Service Compliant Grid Computing;* CRC Press: London, UK, 2009.
9. Wilkinson, B. *Grid Computing Techniques and Applications;* CRC Press: London, UK, 2010.
10. Sharma, P. Grid Computing vs. Cloud Computing. *Int. J. Inf. Comput. Technol.* **2013**, *3*(6), 577–582.

CHAPTER 15

HIGH-PERFORMANCE GRID COMPUTING FOR SCIENCE: A REVIEW

HERU SUSANTO[1,2,*], FANG-YIE LEU[2], and CHING KANG CHEN[3]

[1]The Indonesian Institute of Sciences, Jakarta, Indonesia

[2]Tunghai University, Tunghai, Taiwan

[3]School of Business and Economics, Bandar Seri Begawan, Brunei

*Corresponding author. E-mail: heru.susanto@lipi.go.id

ABSTRACT

The point of having high-performance computing (HPC) is to allow individual nodes to work together in order to solve a problem larger than any one computer can easily solve. These nodes will need to be able to communicate and pass information from one to another interchangeably through variety of computer network. As long as the nodes are connected to the same network, any tasks can still be solved even when all the computers are not located in the same place. It is possible to achieve higher performance computing through a process called HPC which increases computer performance by doing parallel processing to run complicated tasks more efficiently. A high-performance computer or commonly known as super-computer, can be denoted as an output of this process. However, it is a single computer with tens of thousands of processors and is too expensive for every organization to own. HPC is aimed to achieve such performance but this can only partially be done without owning a supercomputer, which is by using multiple computers working together through parallel computing, as mentioned above. Parallel computing is the use of many processors to complete or solve complex computational problems or task. The task is initially divided into different small parts, which will be completed by each of the processors to obtain results faster, and will be transferred to the receiver once every subtasks are completed. Each of the processors is allowed to exchange information with one another, for better performance results. Parallel computers, are actually classified into a few classes based on the tolerance level of parallelism. Parallel computing is often of interest to small and medium sized businesses, as organizations want to achieve supercomputer-performance; although technically it is hardly achievable as a supercomputer is a system that operates at nearly the currently highest operational rate, in which most work at more than a petaflop (thousand trillion) floating-point operations per second. HPC technology focuses on developing parallel processing algorithms and systems by incorporating both administration and parallel computational techniques.

15.1 INTRODUCTION

15.1.1 WHAT IS HIGH-PERFORMANCE COMPUTING (HPC)?

Computing is the process of handling computer technology system, both hardware and software for the purpose of task completion. According to Dayyani and Khayyambashi,[6] high performance refers to the rapidness at which data can be accessed and shared among the set of distributed system. Demands in the speed of data processing system have led to an extensive progress in high-performance computing (HPC), also known as the commodity HPC cluster, a dedicated supercomputer, HPC cloud computing, and grid computing.[8] HPC uses the largest and fastest computer in solving a complex problem, particularly in modern science and engineering. It has provided an exceptional environment, transforming into an indispensable tool in scientific and industrial communities.[23] Sometimes, the collected data are inconsistent and ambiguous, making them even more difficult to be interpreted. HPC systems are notable for such failure as they are usually used for running advanced application programs.[27] HPC assists in solving problems using computer architecture, computer networks, algorithms, programs and electronics, system software, as well as environments in creating adaptable systems.

Now, due to cutting-edge computing technology, HPC is utilized among devices such as smartphones, tablets, laptop, and stand-alone computers. These devices are becoming computational resources based on multi-core processors, which possibly include coprocessor devices that reduce the load on the basic microprocessor circuitry and allow it to perform at optimum speed. Organizations benefit from HPC through its faster solution, better science, informed decisions, more competitive products, and utilizing these things so as to lead to higher profits. HPC represents a tremendous competitive edge in the marketplace because it can provide users the capability to quickly model and then manipulate a product or process to observe the impact of ranges of decisions before they are contrived.

15.1.2   WHAT IS GRID COMPUTING?

According to the information systems concept, grid computing consists of computing infrastructure technologies or machines that gather different computer resources such as storage and processors in order to provide computing support for various applications, turning a computer network into a powerful supercomputer.[22] This computer functions to disseminate information, requiring an environment that provides its ability to share through virtualization not only in a form of servers and CPUs but also utilizing storage, networks, and applications. This virtualization differs inside and outside the enterprise. Inside the business enterprise, virtualization is carried across distributed technologies, platforms, and organizations, emphasizing the social aspect in organizations. Virtualizing externally, however, can be across the internet, making it accessible for the enterprises to scan resources from manufacturers and suppliers. It makes networking and collaborations easier as communication is enhanced through shared information[15] and is often used to complete complicated or tedious mathematical or scientific calculations in standard application areas such as scheduling, security, accounting, and systems management. Grid also enables massively scaled architectures and brings with it a host of the organization as well as technical challenges. No special networking is required for grid, and is not limited to the local-area network.[31]

Grid computing illustrates a simple yet large and powerful virtual computer out of a collection of connected systems of multiple computers, sharing various combinations of resources to complete tasks more efficiently and quickly.

"A computational grid is a hardware and software infrastructure that provides dependable, consistent, pervasive, and inexpensive access to high-end computational capabilities".[22]

15.2   DISCUSSIONS

15.2.1   HOW ARE HPC AND GRID COMPUTING INTERRELATED?

Applying compute cycles to a problem can be done in many ways. According to Eadline,[8] there are four different types of HPC systems. They are known as the commodity HPC cluster, a dedicated supercomputer, HPC cloud computing, and grid computing.[14]

However, among the four, the commodity HPC cluster is substantially the most common and most favored. It is expanded over the entire super-computing market over the last 10 years. This is because the commodity cluster has high performance and a low cost, assembled from interconnected high speed and an off-the-shelf standard. The ordinary commodity cluster had accomplished industry-leading and cost-effectiveness.

The dedicated supercomputer, however, does not offer the commodity bargain but are still produced. Before the commodity HPC cluster, a dedicated supercomputer was the only process to carry a large number of compute cycles to solve a problem. Yet, they are still in use depending on the occasion.

HPC cloud computing is a distributed computing that is built using services scattered over multiple locations, which offers virtualization flexibility, progressive and extensible resources to the end user as a service compared to supercomputers. HPC cloud offers a range of benefits, including elasticity, small start-up, and maintenance costs, and economy of scale. However it also places some layers between the user and hardware that may reduce performance.

Grid computing and HPC cloud computing are complementary but require more control by the person who uses it. It is a more economical way of achieving the processing capabilities of HPC as running an analysis using grid computing is free of charge for the individual researcher once the system is not required to be purchased. Computational grid is similar to a massively parallel supercomputer because its software requires existing computer hardware to function all at once.

Firms take the advantage of the existing hardware for a grid, imposing a huge computing proficiency to their organization. Grid computing is preferred as it saves the organization millions of dollars that costs due to purchasing a supercomputer. Additionally, large-scale commodity cluster is an HPC resource that acts as the building blocks of computational grids.[26] Grid computing and HPC depend on a clearly defined and standard software, but request for updates and other requirements is available. Grid computing is mainly used in academic projects where commodity HPC clusters are connected and shared on a national and international level. Some computational grids span the globe, while others are located within a single organization.

In large HPC systems, input/output forwarding isolates local compute clients, connected by a high bandwidth, low-latency interconnect from the more distant, higher latency parallel file system. Consequently, it is protecting the file system from being harmed by bombarding requests through collecting and merging of requests first and foremost before transmitting them to the file system. From the perspective of the remote file system, this cut down the number of visible clients and requests, hence intensifying the performance. In grid computing environment, these optimizations are also applicable, even though they are on a different scale. While latencies might be much higher, the same discontinuity occurs when an application, running on a local grid resource needs to retrieve data from a remote data store. Similar is the case in large HPC systems, where a substantial amount of simultaneous requests to a remote site might adversely affect the stability and security of the remote file server. This observation is valid both for data staging and wide-area grid file systems.[5]

Grids and HPC are, therefore, generally interrelated in such a way that an organization or institution can provide and access resources at the same time. This is specifically applicable for grid computing as access is "paid" for in-kind by donating access to your own resources.

Upfront development for grid computing can be a long process as different systems are to be dealt with and you cannot control or determine what a specific institute runs on their machines. Using grid and HPC tools on the clouds also allows smaller businesses and individuals to pay only for what they use and allow new business ideas models to come to life without the amount of dollar price tag it can take to put together all the software license and hardware-required solutions.

15.2.2   APPLICATION 1: AUTONOMIC MANAGEMENT

15.2.2.1   WHAT IS AUTONOMIC MANAGEMENT?

Autonomic management targets to demolish the complexity that entangles the management of complex systems, mainly by attaching their parts with self-management facilities.

15.2.2.2   RELATIONSHIP OF AUTONOMIC MANAGEMENT WITH GRID COMPUTING AND HPC

15.2.2.2.1   Relationship of Autonomic Management with Grid Computing

Grid computing, one of the large-scale distributed systems, has its own specific complexity that makes its management a highly troublesome task. The solution to this may possibly be a theoretical one called autonomic computing as it supplies the grid system with the necessary tools that may simplify the system administrator's decisions.[25] Moreover, an alternative way which is to study it as a single entity is utilized for a better understanding since the system involves massive resources that may defect the efficiency of analyzing and implementing policies on each one. This alternative approach involves the technique of self-adaptive with a single entity vision of the grid in order to cater autonomic management and raise the level of dependability.

Grid technologies have enabled the aggregation of geographically distributed resources in the context of a particular application. The network remains an important requirement for any grid application as entities involved in a grid system (such as users, services, and data) need to communicate with each other over a network. The performance of the network must, therefore, be considered when carrying out tasks such as scheduling, migration, or monitoring of jobs. Making use of the network in an efficient and fault-tolerant manner, in the context of such existing research, leads to a significant number of research challenges. One way to address these problems is to make grid middleware incorporate the concept of autonomic systems. Such a change would involve the development of "self-configuring" systems that are able to make decisions autonomously and adapt themselves as the system status changes.

15.2.2.2.2   Relationship of Autonomic Management with HPC

HPC has evolved into becoming a complex and powerful system. Various components include complete computers with onboard CPUs, storage, power supplies, and network interfaces connected to a network (private, public, or the internet) and by a conventional network interface, such as Ethernet, act as roles and build the systems of HPC. The relationship between autonomic management with HPC can be seen through making a connection of how the current HPC systems consume several megawatts of power, which further leads to energy/power efficiency that has to be addressed in a mixture with the requirement of solutions' quality, performance, reliability, and other objectives.

15.2.2.3 AUTONOMIC MANAGEMENT ISSUES IN GRID COMPUTING

The intrinsic complexity of grid systems is difficult to mend by the employment of traditional management techniques. In grid, the system characteristics—heterogeneity, variability, and decentralization—require several and varied perspectives in system management that are not easily adaptable by traditional behavior patterns. Sanchez (2010)[25] states that although these system characteristics may be adapted by adopting an autonomic computing approach, the effects on the grid complexity is unavoidable, namely in its four main areas: self-configuration, self-healing, self-optimization, and self-protection.

15.2.2.3.1 Self-Configuration Issues

The variability and heterogeneity of most grids make resource configuration an inconsiderable amount of problem. Most traditional distributed approaches—cluster computing among other—may very frequently display desirable characteristics that make system configuration nearly an exclusive design problem. Moreover, the process of reconfiguration may usually be performed in an off-line or semi-off-line operation status and simultaneously acquire a specific degree of redesign of the system's structure.

However, the circumstance is entirely different in grids. For one matter, the resources are heterogeneous in nature as well as decentralized and even unforeseeable. Moreover, the resources join and leave the system at a high speed and maybe with variable availability and reliability even though it happens infrequently. Due to these situations, the solution may not be fully fulfilled by a fixed setup as the large-scale-based systems acquire a scalable and adaptable configuration for it to effectively use the available resources provided. Furthermore, if the rate of resource variability is high, it may lead to more complexity of creating an adaptable configuration strategy.

15.2.2.3.2 Self-Healing Issues

Various situations such as resources may appear and disappear at an unpredictable time and rate, and network links may be temporarily or even permanently bothered, and uncontrollable system parts, overloaded by global system administrators, may take place as the aftermath of the natural grid system features. Moreover, all the mentioned situations are considered normal in the eyes of grids' typical behavior environment, even though these variables may have a direct impact on its dependability. However, grids play a vital role as a large sum of resources that fulfill a series of services. Moreover, the typical behavior environments of the grids are difficult to be fault-based categorized as they may be diluted by a high degree of grid cohesion. To conclude, the provision of the high quality of services should be valued in their operation instead of taking into account the nature of its internal resources.

15.2.2.3.3 Self-Optimization Issues

In the point of view of an abstract, optimizing the level of services provided may be entailed by the need to modeling the functionality of the systems as well as being aware and make changes on the limitations of the performance's surroundings. To enjoy the maximization of the system's available resources, a fundamental understanding of the system may aid to develop the advanced management strategies and policies as the basis for performance improvement relies heavily on the understanding of the behavior of the system.

Grid computing features include a large sum of heterogeneous, unpredictable, and non-dedicated resources that communicate during the system's operation process bringing about a whole new level of framework. This may result in an unwilling adaptation of its performance optimization techniques. Furthermore, this may give rise to the main adversity of grid performance optimization, specifically understanding the behavior of the system. The complexity of the large-scale distributed system dilutes the essence of the complicated tasks—such as extraction of the system's behavior pattern and bottleneck identification—as well as its resources' dependencies and deadlocks.

15.2.2.3.4 Self-Protection Issues

The important essence of proactive identification and protection from external attacks is highly focused, given the grids' nature which is distributed, heterogeneous, and decentralized. Due to this, an obligatory initial step is providing shield to each individual resource. This can be executed by the inclusion of traditional and verified techniques to stop it from undisciplined usage as well as other security threats. Unfortunately, these techniques are deemed insufficiently due to the large amount of resource interaction that exists in grid systems, leading to the necessary need for protection gadgets focused solely on the system's global perspectives. However, the grid's optimum complexity makes the aforementioned task heavy, leading to learn the system as well as to analyze the depth of the resources internal and external interactions.

15.2.2.4 AUTONOMIC MANAGEMENT ISSUES IN HPC

Assuming the relationship between autonomic management and HPC on the basis of achieving the main goal of expanding power management at different levels, several autonomic mechanisms are adapted to manage the energy efficiency of HPC through a dynamic cross-layer base in an integrated manner. This adaptation is necessary as current data management approaches are operated on centralized data repositories and are unable to support extreme rates of data generation and distribution scales.[24]

The implication of the modeled energy trade-offs of power management techniques at varying levels as well as the designed model-driven autonomic optimizations and adaptations have revolved on the base of cross-layer approach to producing effective adaptations and optimizations. Moreover, the creation of innovative test beds for the experimental evaluation conduct and the modern-driven platforms such as the Intel SCC multi-core system have been explored thoroughly. Moreover, energy-efficient and thermal-aware autonomic management of HPC workloads is in inclusion of the thorough exploration.

The decomposition of the proposed approach may be effectively implied by following the different stages: (1) categorize relevant data-intensive applications and HPC benchmarks and give a thorough exposure on the matter of power/performance trade-offs to give a definite meaning to the models, (2) develop a set of strategies, application extensions, and modes of usage so as to be accepted by the model-driven optimizations and adaptations, (3) permits cross-layer interactions and integrate them with the runtime system, and lastly (4) adapt simulation into the exploration of the suggested approach at a larger scope as well as taking into account the nonstandard hardware configurations, hardware, and software codesign and determine how these components have an impact toward the applications and system.

Furthermore, the main challenges of the specific cross-layer tactic can be summed as: (1) understanding the application models and maintaining the application to help facilitate in performing effective optimizations as well as adaptations at varying measurements; (2) studying the applications of these adaptations being implied toward other dimensions which may include performance, data management, and the methods of using the different stages of memory hierarchy; and (3) comprehending the control plane and mechanisms usage in order to implement and fulfill such a management approach.[24]

15.2.3 APPLICATION OF BOTH HPC AND GRID COMPUTING IN FINANCE SECTOR

15.2.3.1 ROLE OF HPC IN FINANCE SECTOR

The impression of the financial turmoil on information and communication technology (ICT) has been quick and boundless, which resulted in a general reduction in investment and clearly created the idea that all the priorities need to be on how ICT should be the core focus of the business to recover its profitability and support the changes and streamlining. On the contrary, the ICT departments are developing hard-pressed by the conflicting demands of cutting costs set against the clear idea for an improved reactiveness and immediate response times to the demand of the business.[21]

HPC has been quite prominent in the financial sector for years even if it was too bulky or highly complicated to use. At the present moment, HPC is used to solve complicated and advanced computation problems.

HPC is mainly utilized in financial services for regulatory compliance, pricing, pre-trade risk analysis, and future trends to meet with. Financial sector clearly opts for HPC mainly because of the fact that the ability of it to handle a massive number of data at a very high speed as well as the ability to solve complex computation.

Garg et al.[10] stated an example whereby if an investor would like to know on how to make a decision whether to purchase a stock at a future date and want to retrieve any information at a particular price based on some basic information visible to public such as Yahoo! Finance. This type of future investment is known as an option, and this requires knowledge of the current stock price and the chances of when to exercise the stock for the profit. As there are plenty of algorithms to price an option which creates this problem called an option pricing problem in finance. If the investors are interested in pricing an option, they would need to have the working knowledge of these algorithms to help them make a great and informative decision on computing the option prices. The algorithms used in the option pricing problem are computationally intensive and require parallel processing to obtain results in real time. The option pricing problem uses the algorithms which are very complex for an investor to understand as it is not easy to understand the algorithms with no backgrounds of it. The option pricing problems fall under the category of HPC application and have successfully done several works in this field.

Joseph and Fellenstein[16] stated, "A financial organizations processing wealth management application collaborates with the different departments for more computational power and software modeling applications." They also mentioned that it equips a number of computing resources, which results in higher performance with higher and real-time executions of the tasks and prompt access to higher data storage while managing complex data transfer which obviously resulted in higher number of consumer's satisfaction with a good amount of time.

One of the very famous scenarios that made the financial sector usher its way to using HPC was during the financial crisis of 2007–2008, which was hit badly due to the miscalculation as the traditional computer technology were not reliable to handle such complex and heavy computations. Since then, HPC has been shining around the financial sector.

As for now, stated by Stanoevska-Slabeva et al.,[29] companies need to survive and develop competitive advantage in a dynamic and in an unstable environment of global competition and accelerated business change.

There is no doubt that the companies are always under peer pressure to simultaneously grow revenue and market share while trying their very best to reduce the financial cost incurred. They have also mentioned that to meet these requirements, companies will have to evolve in the three most major trends that will have an impact on the companies wants using the information technology (IT) support:

1. Surviving towards high awareness on what the latest trend is
2. Globalization of activities to be able to take advantage of opportunities provided by a global economy
3. Increased maneuverability

15.2.3.2 ROLE OF GRID COMPUTING IN FINANCE SECTOR

Bingxiang and Lihua[4] stated, "Grid computing is an advanced computing infrastructure, and its most prominent feature is heterogeneous, dynamic nature and shared nature."

The application of the grid technology can be adequately strengthening the financial system, the ability to process data, different technical aspects of resolving the present problem, thereby cutting the financial cost of doing business. According to Bingxiang and Lihua,[4] many international companies have well established in their respective financial grid system to realize the demand, disposal of resources, and efficient cost savings.

They also mentioned that the use of this technology, which can adequately strengthen the financial system, the ability to process the data, finding solutions for current complex computation problems may benefit the firms to cut down on the financial cost incurred by doing business. Moreover, the adaptation of the grid technology by many financial companies have settled their respective financial grid system to apprehend their demand on deploying of the resources, on improving resources utilization and efficiency and saving cost.

The financial system needs to build a virtual grid focal point in the system in which it can provide the means of computing, information service, and other resources to be shared. The problem arises when the intensity of the calculation to carry out is lagging behind in the financial sector, which will affect the product innovation, pricing, loan analysis, and utilizing the resource fully to achieve the business's objective as majorly and also when the rather expensive machines are not in use or are in idle state, which may incur cost.[32]

According to Zhang,[32] the grid system used can be divided into resource layer, middleware layer in which it acts as a middleman between the software and the database, the web environment layer having four financial uses. First, the resource layer, which includes the grid systems' main groundwork, the famous grid nodes, and broadband network systems, in which the node consists of different resources such as supercomputers, application software, databases, and cluster systems. The main objective is the hardware and software resources to provide the access the platform and to control management.

15.2.3.3 FUTURE OF HPC AND GRID COMPUTING

Bingxiang and Lihua[4] stated, "However, with the further expansions of financial services, the traditional IT infrastructure can no longer meet the information needs of social development, we must create a new open grid structure to achieve the financial 'resource sharing co-processing' in order to keep up with the financial international trend."

In relevance to what International Data Corporation (2015) found out that the business firms are changing to HPC rapidly because of the advantages as well as the reduction in cost. Even though the recent trend states that there is a great potential growth for the HPC, still there are challenges like having a difficulty to find enough data center space and power. Moreover, the bandwidth limitations play a huge role in moving around large data sets with these geographically dispersed networks with rather high reliability and high bandwidth interconnectivity.

In future, it is predicted that the cloud computing may expand to its fullest as it has a clear advantage over HPC. In contempt of the similarities between HPC and cloud computing, they do differ in many known ways. Both are driven by cost, reliability, and energy efficiency imperatives; the commercial cloud will require more continuous service even when there is some failure.[11]

Shi et al.[28] stated that "With the advances of underlying technologies for cloud computing, it is increasingly gaining applications in the financial service industry."

Furthermore, HPC is not only too bulky but consumes a lot of space compared to recent innovation, cloud computing. With this being in trend helps even more for the business to be capable of scaling up or down rapidly as well as the availability.

For instance, NASDAQ OMX Group had disclosed around the year 2012 the launch of a new cloud computing named, FinQloud. This was mechanized by Amazon Web Services and is only particularly made for the financial service industry. This clouding platform caters regulators, brokers–dealers, and advisory services to exchange all within the global framework.

Another example would be "Aneka," an enterprise cloud computing solution, which harnesses the power of computing resources by relying on private and public clouds and delivers to users the desired quality of service. It is quite adaptable which supports multiples of programming protocols that make Aneka quite different. It is mainly designed for finance applications to computational science.[30]

15.2.4   APPLICATION 3: SERVICE-LEVEL SYSTEM MANAGEMENT

In this section, we will use research that has been conducted as evidence to back up our claim that grid computing and HPC are used in service-level system management. Some examples of service-level system management are healthcare, e-science, and school. Additionally, we will also cover some of the drawbacks of grid computing and HPC and how to overcome the challenges.[1]

15.2.4.1   HEALTHCARE

Grid computing and HPC play an important role in healthcare. Karthikeyan and Manjula[17] stated that grid computing enables doctors to make medical decisions, easier transfer and collaboration of information, as well as improve medication management. They further added that grid computing may offer more advanced services such as telemedicine, daily monitoring, and exchanging information between two or more parties.

Electronic services (e-services) such as storing the patients' record can be improved with the help of grid computing.[19] The fact that patients' records were in paper makes it environment-unfriendly because to edit the record, more papers have to be attached to the file, which is time-consuming and the records could be lost.[19] These problems can be addressed if grid computing is used instead. In the same piece of journal article, Liu and Zhu mentioned that the resource grid layer which is where grid computing is used have two primary functions. They are to help providers and clients connect and to build a reliable setting in order for the business operates and transaction to take place over the internet through grid computing.

According to Guerrero et al.,[13] HPC is vital to capture more biological knowledge. Their study was focused on how HPC can benefit bioinformatics. They stated that having the most efficient performance is favorable but at the same time, other organizations should focus more on cutting the cost to build an HPC. They found out that volunteer computing is not the solution for all remedies because it relies on the severity of the problem and compatible materials that are given during a project. Overall, HPC is used mainly to gather knowledge with less delay due to the higher GPU which the computers use.

15.2.4.2   E-SCIENCE

E-science is a branch of science that requires the usage of significant amounts of computing resources and massive data sets in order to perform scientific inquiry. The data produced may need the experts' scrutiny.[20] Mustafee[20] also explained that grid computing is an asset for e-science due to its wide applications in various fields. According to him, some of the projects which rely on grid computing were the Earth System Grid, which focal point was climatology; Large Hadron Collider, which focuses on particle physics and creating a simulation for earthquake engineering. However, grid computing is costly to produce yet it is regarded as an investment because grid computing is used to solve bigger problems.[20]

Barati et al.[2] proposed an algorithm in their research which will minimize the failure of grid computing so that the same algorithm can be applied in their e-science projects for resource management. They tried to minimize the occurrence of error when one of the agents failed to complete the resources chain by defining a new agent called task agent. The function of the task agent is to create subagents so that they can communicate with the resource agent in order to check the success or failure of the resource agent. If there is a breakdown in the resource agent, the subagent will find another resource agent which has the required data and maintain connectivity with the new resource agent. They found out that there is an inverse relationship between increasing the number of resource types and the success percentage of the data transferred. However, they are still willing to use their algorithm in the near future probably with more advancement and hopefully overcome the inverse relationship between success percentage and the number of resource types.

HPC is also used in the science field as well as other different areas and this is a proven fact by Benedict.[3] In his survey, he found that although HPC-based cloud applications are used in various fields of work, HPC-based cloud applications still faced some obstacles such as resource outages, service-level agreement violations and the lack of adequate performance analysis tools. However, in his research, he came up with a few solutions to tackle these problems and hopefully some of them are being implemented as it will bring greater advantage to all parties concern.

15.2.4.3 SCHOOL

Given that schools usually take up a bigger space, there is no doubt that grid computing can be utilized as it can be linked to the internet or low-speed networks. Grid computing is particularly useful as a school tends to have a minimum of hundreds students who have access to the school's network, which may cause low speed of the internet. Li et al.[18] explained the execution process if a person wishes to operate the learning platform. First, the learner will have to enter the grid portal and key in the user's login name and password. The system will provide a list of resources available along with the status if the login is acknowledged. The circulation of computation or data across other computers in the same grid is being assigned by a broker. GridFTP (file transfer protocol) is used to access the computer's resource for the distribution of data, whereas, for the resource which has high-speed performance, Replica Location Service is used instead. Thus, the students can benefit from grid computing.

Another study conducted by Goldsborough,[12] however, proves that grid computing is not necessarily the form of file sharing that is popular, it is instead cloud computing. The reasons why cloud computing is more favorable compared to grid computing are that it does not require high-intensity processing power and space to contain valuable information, it is cost-friendly as nearly all the time it is free of charge, and is less time-consuming. It brings benefit to not just businesses but also to other organizations such as schools. In addition to the many advantages cloud computing offers, perhaps the most important one is cloud backup services. We never know when our technology will fail us due to the virus, our own clumsiness, blackouts, and so on. Cloud computing such as Google Drive is very useful to act as a "safe house" and it is very user-friendly. Hence, grid computing seems to be on the losing end against cloud computing as it seems that cloud computing is higher in demand.

Eurich et al. validated that one of the benefits of using HPC is that it can be used to make business models.[9] They discussed the pros and cons that HPC provide for business models which can be utilized by higher-education-based HPC centers' directors, schools, and government to design an acceptable business model. They also predicted that the pricing strategies and the design of the business models will be more relevant in the future. The decision makers have to do research on their consumers, investigate the national e-infrastructure to keep up with the current trends.

15.2.4.4 DRAWBACKS

Despite the advantages that HPC possess, there are a few challenges to make these supercomputers. According to Dongarra and Van Der Steen,[7] supercomputers could face some issues such as standardization, developing more mathematical models and algorithms. In the same piece of journal article, they explained that while making the software, standardization is the minimal condition. However, these problems can be tackled if there is a group of experts in various fields working together to make a supercomputer.

According to Mustafee,[20] there is a possibility that cloud computing might take over grid computing because grid computing is not dynamically scalable. Due to grid computing not being dynamically scalable, e-science applications rely more on cloud computing. Goldsborough[12] also points out many advantages that cloud computing have—some of them are absent for grid computing. Hence, in order to make grid computing favorable in the future, further research needs to be conducted to find out ways to promote grid computing or find the specific field where grid computing is considered the best solution to their problem.

15.3 CONCLUSION

When Ian Foster, Carl Kesselman, and Steve Tuecke brought up the idea of grid computing into reality, they also rolled out the idea of distributed computing, object-oriented programming and also the web service. It is in the early 1990s, the term grid computing was introduced and is used as a metaphor for making computer power as easy to access as an electronic power grid.

Grid is defined as sharing resources between multiple parties, for example, the government, universities, computing centers, which have a need to do a joint work on a project for a period of time. It not only concentrates on file exchange but also allows an open access to computers, software, data, and other resources.

The access to high-end computational resources with the hardware and software facility needs to be dependable, consistent, pervasive, and inexpensive. It is also expected that the usage of grid computing in our everyday life can serve multiple colonies such as government, private sectors, and so forth. Hence, some national-level problems that are correlated with the government are been solved through the use of nation's superfast computers and data archives.

There are many areas in service-level management that require the use of grid computing. Some examples given are healthcare, e-science, and school. The use of grid computing in these area help to reduce hassle working environment. Some hassle working environment such as bureaucratic paperwork and the time needed for completing a specific task is reduced. The risk of loss of record is also reduced and replaced by those real-time tracking and recording.

In conclusion, grid computing is important in many areas. Government and important IT supplier such as automated interchange of technical information plays an important role in enhancing and implementing these technologies. However, first, it had to be integrated with several other computer systems such as cloud computing, distributed computing, object-oriented programming, and also the web service in order to make them work more effectively and efficiently. Once both effectiveness and efficiency are achieved in the long run, productivity can increase and operating costs can be reduced which can eventually help in saving and reducing the budget of an institute or an organization.

KEYWORDS

- **grid computing**
- **high-performance computing**
- **information and communication technology**
- **local area network**
- **parallel computers**

REFERENCES

1. Aldinucci, M.; Campa, S.; Danelutto, M.; Vanneschi, M. *Behavioral Skeletons in GCM, Autonomic Management of Grid Components*, University of Pisa: Italy, 2008, pp 56.
2. Barati, M.; Lotfi, S.; Rahmati, A. A Fault Tolerance Algorithm for Resource Discovery in Semantic Grid Computing Using Task Agents. *J. Software Eng. Appl.* **2014**, *7*(4), 256–263. http://search.proquest.com/docview/1524713469? accountid=9765
3. Benedict, S. Performance Issues and Performance Analysis Tools for HPC Cloud Applications: A Survey. *Computing. Arc. Inf. Numer. Comput.* **2013**, *95*(2), 89–108. http://search.proquest.com.ezproxy.ubd.edu.bn/docview/1283134915/DC779 DBE04CE41DFPQ/13? accountid=9765
4. Bingxiang, G.; Lihua, J. Research on Grid Technology Applied in the China Financial Sector. *Lect. Notes Electr. Eng. Future Commun. Comput. Control Manage.* **2012**, 1–6. http://link.springer.com/chapter/10.1007/978-3-642-27,314-8_1#page-1
5. Cope, J.; Iskra, K.; Kimpe, D.; Ross, R. I/O Forwarding in Grids. State of the Art in Scientific and Parallel Computing, 2010. http://www.mcs.anl.gov/research/projects/iofsl/pubs/para10-extabs.pdf
6. Dayyani, S.; Khayyambashi, M. R. A Comparative Study of Replication Techniques in Grid Computing Systems. *Int. J. Comput. Sci. Inf. Secur.* **2013**, *11*(9), 64–73. http://search.proquest.com/docview/1468454387? accountid=9765
7. Dongarra, J. J.; van der Steen, A. J. High-Performance Computing Systems: Status and Outlook. *Acta Numer.* **2012**, *21*, 379–474. http://search.proquest.com.ezproxy.ubd.edu.bn/docview/1002683861/CF6569B03AAD41E4PQ/2? accountid=9765
8. Eadline, D. High Performance Computing For Dummies, 2009. http://hpc.fs.uni-lj.si/sites/default/files/HPC_for_dummies.pdf
9. Eurich, M.; Calleja, P.; Boutellier, R. Business Models of High Performance Computing Centres in Higher Education in Europe. *J. Comput. Higher Educ.* **2013**, *25*(3), 166–181. http://search.proquest.com.ezproxy.ubd.edu.bn/docview/1448980286/CF31 C07F1D304E14PQ/8? accountid=9765
10. Garg, S. K.; Sharma, B.; Calheiros, R. N.; Thulasiram, R. K.; Thulasiraman, P.; Buyya, R. Financial Application as a Software Service on Cloud. *Commun. Comput. Inf. Sci. Contemp. Comput.* **2012**, 141–151.

11. Geist, A.; Reed, D. A. A Survey of High-Performance Computing Scaling Challenges. *Int. J. HPC Appl.* **2015**, *3*. http://hpc.sagepub.com.ezproxy.ubd.edu.bn/content/early/2015/08/05/1094342015597083.full.pdf html

12. Goldsborough, R. Computing in the Cloud. *Tech Directions*, **2010**, *70*(5), 14. http://search.proquest.com/docview/819261026? accountid=9765

13. Guerrero, G. D.; Imbernón, B.; Pérez-Sánchez, H.; Sanz, F.; García, J.; M.; Cecilia, J. M. A Performance/Cost Evaluation for a GPU-Based Drug Discovery Application on Volunteer Computing. *BioMed Res. Int.* **2014**. http://search.proquest.com.ezproxy.ubd.edu.bn/docview/1552819722/BAB0EEFD32ED4A77PQ/9? accountid=9765

14. Russell, J. IDC: The Changing Face of HPC. **2015**. http://www.hpcwire.com/2015/07/16/idc-the-changing-face-of-hpc/ (accessed Feb 25, 2016).

15. Jacob, B.; Brown, M.; Fukui, K.; Trivedi, N. Introduction to Grid Computing. IBM Redbooks, 2005, pp 3–6. https://www.redbooks.ibm.com/redbooks/pdfs/sg246778.pdf

16. Joseph, J.; Fellenstein, C. *Grid Computing. Upper Saddle River*, 2004, Prentice Hall Professional Technical Reference: NJ.

17. Karthikeyan, P, Manjula. K. A. Business Applications of Grid Computing: A Review. *Nat. J. Adv. Comput. Manage.* **2010**, *1*(2). http://search.proquest.com/docview/1765404098? accountid=9765

18. Li, K.; Tsai, Y.; Tsai, C. Toward Development of Distance Learning Environment in the Grid. *Int. J. Distance Educ. Technol.* **2008**, *6*(3), 45–57. http://search.proquest.com/docview/201699518? accountid=9765

19. Liu, L.; Zhu, D. An Integrated e-Service Model for Electronic Medical Records. *Inf. Syst. eBusiness Manage.* **2013**, *11*(1), 161–183. http://http://search.proquest.com.ezproxy.ubd.edu.bn/docview/1283523065/6CD5F6439BED4A8APQ/6? accountid=9765

20. Mustafee, N. Exploiting Grid Computing, Desktop Grids and Cloud Computing for e-Science. *Transforming Gov.: People, Process Policy* **2010**, *4*(4), 288–298. http://search.proquest.com.ezproxy.ubd.edu.bn/docview/761437812/D40BAF0295584220PQ/1? accountid=9765

21. Nicoletti, B. *Cloud Computing in Financial Services*; Palgrave Macmillan: New York, 2013.

22. Oesterle, F.; Ostermann, S.; Prodan, R.; Mayr, G. J. Experiences with Distributed Computing for Meteorological Applications: Grid Computing and Cloud Computing. *Geosci. Model Dev.* **2015**, 2067–2078. DOI: 10.5194/gmd-8-2067-2015.

23. Ranilla, J.; Garzón, E.; Vigo-Aguiar, J. High Performance Computing: an Essential Tool for Science and Engineering Breakthroughs. *J. Supercomput.* **2014**, 511–513. DOI: 10.1007/s11227-014-1279-6.

24. Rodero, I.; Parashar, M. Cross-Layer Application-Aware Power/Energy Management for Extreme Scale Science. Rutgers University: Rutgers Discovery Informatics Institute and NSF Cloud and Autonomic Computing Centre, 2012.

25. Sanchez, J. M. Global Behavior Modeling, A New Approach To Grid Autonomic Management, Ph.D. Thesis, Technical University of Madrid: Madrid, 2010, 16, 65–70.

26. Selvi, S. Resource Management System for Computational Grid. *Int. J. Multidiscip. Approach Studies* **2015**, *2*(4), 1–4.

27. Shawky, D. Scalable Approach to Failure Analysis of High-Performance Computing Systems. *ETRI J.* **2014**, 36(6), 1023–1031. DOI: 10.4218/etrij.14.0113.1133.
28. Shi, A.; Xia, Y.; Zhan, H. Applying Cloud Computing in Financial Service Industry [Abstract]. 2010 International Conference on Intelligent Control and Information Processing, 2010.
29. Stanoevska-Slabeva, K.; Wozniak, T.; Ristol, S. Grid and Cloud Computing: A Business Perspective on Technology and Applications. Springer: Heidelberg, 2010.
30. Vecchiola, C.; Pandey, S.; Buyya, R. High-Performance Cloud Computing: A View of Scientific Applications [Abstract]. 2009. 10th International Symposium on Pervasive Systems, Algorithms, and Networks, 2009.
31. Vile, A.; Liddle, J. The Savvy Guide to HPC, Grid, Data Grid, Virtualisation and Cloud Computing, 2008.
32. Zhang, Y. *Future Communication, Computing, Control and Management* 2014; Vol. 2, Springer Berlin: Berlin.

INDEX

Printed in the United States
by Baker & Taylor Publisher Services